L. F. M. van Zutphen · M. van der Meer (Eds.) Welfare Aspects of Transgenic Animals

Springer
Berlin
Heidelberg
New York
Barcelona
Budapest
Hong Kong
London
Milan
Paris
Santa Clara
Singapore
Tokyo

L. F. M. van Zutphen · M. van der Meer (Eds.)

Welfare Aspects of Transgenic Animals

Proceedings EC-Workshop of October 30, 1995

With 2 Figures and 6 Tables

Springer

Prof. Dr. L. F. M. van Zutphen
M. van der Meer, MSc.

Utrecht University
Faculty of Veterinary Medicine
Department of Laboratory Animal Science
P.O. Box 80166
NL-3508 TD Utrecht
Netherlands

ISBN-13:978-3-540-61839-3 e-ISBN-13:978-3-642-60589-5
DOI: 10.1007/978-3-642-60589-5

Library of Congress Cataloging-in-Publication Data applied for

Welfare aspects of transgenic animals: Proceedings EC-Workshop of October 30, 1995 / ed. L. F. M. van Zutphen and M. van der Meer. Department of Laboratory Animal Science, Utrecht University, The Netherlands. - Berlin; Heidelberg; New York; Barcelona; Budapest; Hong Kong; London; Milan; Paris; Santa Clara; Singapur; Tokyo: Springer, 1997 ISBN-13: 978-3-540-61839-3 NE: Zutphen, Lambertus F. M. van [Hrsg.]; EC-Workshop <1995, Utrecht-; Universiteit <Utrecht> / Proefdierkunde

Production: PRO EDIT GmbH, D-69126 Heidelberg
camera ready by author
Cover Design: design & production GmbH, D-69121 Heidelberg

SPIN: 10556257 27/3136-5 4 3 2 1 0 - Printed on acid-free paper

Preface

The quality of animal models used in biomedical research has increased substantially since the first successful microinjection of foreign DNA fragments into a mouse zygote in the early 1980s. New well-defined animal models, developed either by microinjection techniques or by homologous recombination in embryonic stem (ES) cells, have become widely available and several classical animal models have now been replaced by these transgenic models. Transgenic animals also fulfill increasingly important roles in other fields, like livestock production or in biopharming.

The potential benefits of this technology are overwhelming. It is not surprising that many research groups have rushed to jump on this wagon of scientific progress.

However, no matter how important these developments are, they can not relieve us of the moral obligations towards the animals. It is generally recognized that animal use is allowed only if the harmful effects are outweighed by the potential benefits of the experimental results. But what are the harmful effects? What is the impact of transgenesis on the well-being of animals?

In order to gain more insight into this aspect of transgenesis the European Commission (DGXII) has funded a 12-month project with the aim of collecting information on the production and use of transgenic animals in EU Member States and, more specifically, to obtain information as to what extent specific welfare problems may arise and whether or not studies are being performed on the incidence of welfare problems.

The results of this study will be published in a separate report by Miriam van der Meer (Department of Laboratory Animal Science, Utrecht University). This project also included the organization of an European Commission workshop on welfare aspects of transgenic animals.

This workshop is organized in the framework of the European Commission's Biotechnology Programme. Part of this programme is the funding of studies on the ethical, social and economic effects of biotechnology.

The question may rise as to why special attention should be paid to the well-being of animals in relation to transgenesis?

There are probably several answers to this question. One argument could be that transgenesis is a relatively new, but widely used technique, involving a great number of animals. Also, the technique is emerging in a period of time in which the position of animals in relation to man is being reconsidered, or at least is an issue of increasing debate. And although the views in this debate are diverse, there is one mutually accepted theme: the concern that everything possible should be done to ensure the well-being of animals or to avoid them suffering unduly.

But there is more. Because the technique of transgenesis has enabled us to exchange genetic material between species, it also has become a major ethical issue.

The question, however, remains whether we can expect that the well-being of animals produced by transgenesis is more at risk than of animals produced by traditional breeding methods. For instance, are there reasons to expect that a transgenic animal experiences more discomfort than an animal in which a similar gene has been introduced by producing a congenic line.

It can be argued that there is indeed an increased chance of welfare problems in animals produced by microinjection techniques when compared to the process of producing mutant animals by traditional breeding methods. These problems go beyond those of the direct effect of the mutant gene and can be attributed to the potential disturbance of the balanced genotype by insertional mutations. The question as to what extent this type of insertional mutation is responsible for welfare problems still remains to be answered.

The proceedings of this workshop consist of the nine papers presented by invited speakers and a brief summary of the general discussion. The first four chapters of these proceedings contain contributions in which transgenic techniques and applications of these techniques are the central themes. In each of these chapters the author also gives a personal view on animal welfare considerations. Chapters 5 and 6 deal with general and practical aspects of assessing animal welfare of transgenic animals. Chapter 7 provides a brief overview of the results of the above mentioned EC project. The ethical and legal aspects of producing and using transgenic animals are presented in chapters 8 and 9, respectively. The final chapter gives a brief summary of the general discussion.

Various views as to the implications of transgenesis on the welfare of animals are presented in these proceedings. In this respect there is certainly no consensus yet. Nonetheless there seemed to exist general agreement at the workshop that the welfare of animals is an important aspect to be considered when starting the production or use of transgenic animals. It was, however, recognized that there is still a lack of knowledge as to the degree of animal's suffering due to transgenic techniques. The conclusion must be that there is an urgent need for more scientifically based information in this field.

The editors wish to thank Mrs. Marianne Albers for streamlining the manuscripts into a uniform style and Dr. Donald Straughan for his invaluable linguistic advices. The European Commission (DG XII) is acknowledged for funding the workshop.

M. van der Meer
L.F.M. van Zutphen

Utrecht, September, 1996

Contents

List of Contributors

Prof. D.M. Broom
Department of Clinical Veterinary Medicine, University of Cambridge, Madingley Road, Cambridge CB3 OES, United Kingdom

Dr. K. Bürki
Preclinical Research, Sandoz Pharma Ltd., CH-4002 Basel, Switzerland

Dr. P. Costa
Instituto di Ricerche di Biologia Molecolare P. Angeletti S.p.A., Via Pontina KM 30.600, 00040 Pomezia, Rome, Italy

Dr. B. Forsman
Centre for Research Ethics, The Royal Society of Arts and Sciences in Gothenburg, Brogatan 4, S-413-01 Gothenburg, Sweden

Drs. P. de Greeve
Department of Animal Experimentation, Veterinary Public Health Inspectorate, Ministry of Public Health, Welfare and Sport, P.O. Box 5406, 2280 HK Rijswijk, The Netherlands

Dr. A.K. Hansen
Department of Experimental Medicine, University of Copenhagen & National University Hospital, The Panum Institute, Blegdamsvej 3B, DK-2200 Copenhagen N, Denmark

Drs. W. de Leeuw
Department of Animal Experimentation, Veterinary Public Health Inspectorate, Ministry of Public Health, Welfare and Sport, P.O. Box 5406, 2280 HK Rijswijk, The Netherlands

Drs. M. van der Meer
Department of Laboratory Animal Science, Faculty of Veterinary Medicine, Utrecht University, P.O. Box 80.166, 3508 TD Utrecht, The Netherlands

Dr. W. Müller
Institute for Genetics, University of Cologne, Weyertal 121, D-50931 Cologne, Germany

Dr. T. Poole
UFAW (Universities Federation for Animal Welfare), 8 Hamilton Close, South Mimms, Potters Bar, Herts EN6 3QD, England

Dr. P. Sandøe
Bioethical Research Group, University of Copenhagen, Dept. of Education, Philosophy and Rhetorics, Njalsgade 80, DK-2300 Copenhagen S., Denmark

Dr. J.S. Verbeek
University Hospital, G04.614, Department of Immunology, Heidelberglaan 100, 3584 CX Utrecht, The Netherlands

Dr. R. Wanke
Institut für Tierpathologie, Ludwig-Maximilians-Universität, Veterinärstrasse 13, D-80539 München, Germany

Prof. Dr. E. Wolf
Lehrstuhl für Molekulare Tierzucht und Haustiergenetik, Ludwig-Maximilians-Universität München, Genzentrum, Würmtalstrasse 221, D-81375 München, Germany

Prof. Dr. L.F.M. van Zutphen
Department of Laboratory Animal Science, Faculty of Veterinary Medicine, P.O. Box 80.166, 3508 TD Utrecht, The Netherlands

Scientific Applications of Transgenic Mouse Models

J. S. Verbeek

Summary

The fast growing knowledge about the complex biology of higher eukaryotic systems demanded new experimental models. Transgenesis of mammals is one of the most fruitful techniques to create these models.

The transgenic mouse model has evolved to a full fledged genetic system that allows all types of genetic modifications known already for a long time from genetic systems in lower organism like bacteria, yeast and Drosophila. On the one hand we are able to add (modified) genes to the resident genetic material, on the other hand we can knock-out or alter any gene of the resident genetic material. The mouse combines characteristics such as low cost, easy handling, and short generation time, with the availability of a large number of genetically and immunologically well-characterized inbred strains. In addition, the mouse shares with man most metabolic pathways and physiological processes. In conclusion this makes the transgenic mouse very useful in fundamental and medical science.

In this paper, the power of the technique of transgenesis in the mouse will be demonstrated by a selection of recent examples from our own laboratory. The examples include transgenic mouse models generated to study the T lymphocyte specific transcription factor, T cell factor 1 (TCF-1); Xeroderma Pigmentosa complementation group A correcting gene (XPA) involved in DNA excision repair; receptors for the Fc part of immunoglobulins of the IgG class (FcγR) and Islet amyloid polypeptide produced by the ß cells of the Islets of Langerhans in the pancreas.

The following aspects related to welfare and ethics will be discussed briefly: lack of control of the integration site of the transgene; the effects of heterologous transgenes on the identity of the transgenic animal; the influence on the well-being of the transgenic animal by the expression of the transgene; and, the risk to the environment.

Introduction

One of the best ways of establishing the function of genes and their products is to observe what happens when the genes are mutated. This is the simple principle of classical genetic analysis. In the past, geneticists and molecular biologists relied on spontaneous or induced mutations and genetic polymorphism to generate genetic maps and to identify genes. However, classical genetics has its limitations in the analysis of the functions of the 40-100,000 genes in the complex diploid genome of mammals as: only mutations that produce a readily detectable phenotype can be studied; in diploid organisms many mutations can only be detected after appropriate breeding strategies to generate homozygotes for the mutation; and many mutations are lethal in homozygotes.

With the advent of recombinant DNA techniques, any gene can, in principle, be isolated, multiplied to large quantities and physically analyzed. In order to study its function the gene can be introduced in a living cell by gene transfer techniques: the concept of reverse genetics. This creates the following experimental options: all possible mutations can be studied by simply changing the cloned DNA before introducing it into the cell (modulation-of-function); the gene can be transferred to a cell which does not normally have the gene (gain-of-function); a gene, highly homologous to an endogenous genomic gene, can be introduced in a modified form followed by homologous recombination with the endogenous gene resulting in the specific modification (change-of-function) or inactivation (loss-of-function) of that endogenous gene.

This approach has shown to be very helpful in the identification and analysis of several genes. Moreover *in vitro* gene-expression systems can be used for the large scale production of (pharmaceutical) proteins. However, *in vitro* systems have their limitations for the following reasons: gene expression is often cell-type specific; not all different cell types can be cultured *in vitro*; and, cell lines do not retain their original tissue-specific characteristics completely. Furthermore, in complex interactive systems like body plan formation during embryogenesis or the response of the immune system, it is impossible to assess the role of proteins that are involved in a simple cell culture system.

The definitive way of analysing gene expression therefore, is to transfer the genes back into an intact animal and to follow the expression during development and differentiation. This is now possible in a number of different organisms: fruit fly, mouse, rat, cattle, sheep, pig, goat, rabbit, fish and chicken.

In the mouse there are two general approaches. The first method, developed in the beginning of the eighties is to introduce the DNA directly into cells of one of the pre-implantation stages of the embryo either by micro-injection of fertilized oocytes or by infection of later stages with retroviruses. If the construct integrates stably in the genome, then there is a good chance that it will be inherited by most if not all cells of the developing animal, the germ cells included (germ line integration). Introduction of new genetic information in the germ-line of organisms is called 'transgenesis' resulting in 'transgenic animals' that can transmit their newly acquired genetic information to their offspring in a Mendelian fashion.

The second, recently developed, approach, is to introduce genetically manipulated pluripotent embryonic stem cells into a late pre-implantation stage embryo. By their pluripotency, the introduced cells are able to take part in the embryonic development and to contribute to most if not all different tissues of the developing animal that therefore becomes a chimaera. When the introduced embryonic stem cells are transfected with a DNA construct and contribute to the formation of the germ line the resulting mouse is a transgenic chimaera that may produce fully transgenic progeny. The major advantage of this procedure is that the gene transfer takes place in the embryonic stem cells, that are available now (in the mouse system) as continuous cell lines. This enables us to introduce precise defined mutations in endogenous genes by gene targeting via homologous recombination. The frequency of homologous recombination between a transferred DNA construct and homologous endogenous genomic sequences is very low compared to the random integration of the DNA construct in the genome. *In vitro*, however, it is possible to select for cells in which homologous recombination took place. After selection these cells can be used to generate transgenic chimaeras.

More recently, by using the bacteriophage P1-derived *Cre* recombinase together with short phage sequences, the *loxP* sites, it was shown that even conditional gene targeting can be performed in living mice. Conditional gene targeting in a specific cell type allows the possibility of circumventing the lethality of an otherwise fatal mutation. This approach will be discussed by Werner Müller in the next chapter.

Applications

The results obtained with transgenic mouse models come mainly from the following 5 research areas: the molecular biology of gene regulation, immunology, developmental biology, oncology and disease models. (For review see: Palmiter and Bristol 1986; Jaenisch 1988; Hanahan; 1989). The selection of recent examples from our own laboratory that will be discussed in this chapter are related respectively to the molecular biology of gene regulation, oncology, immunology and disease models.

The Molecular Biology of Gene Regulation

To regulate gene expression at the level of transcription, the coding sequences of a gene are always flanked by regulatory sequences. The regulatory regions of many different genes have been identified via transgenesis with gene constructs harbouring different fragments of gene-flanking DNA regions. If a transgene, including its regulatory regions, is expressed at all, it is generally expressed in the correct developmental and tissue-specific fashion. Very frequently expression levels are close to the levels of expression of the endogenous cognate genes. This indicates that the site of integration is not pivotal in the regulation of expression, and that the structural organization of the chromosomal DNA is apparently flexible enough to incorporate and allow correct expression of large pieces of exogenous DNA. Even if the transgene has been derived from another species, the developmental and tissue-specific expression is maintained. This indicates that the expression is regulated by transacting factors which have been sufficiently conserved in evolution in order to recognize and act upon regulatory sequences present in DNA from different species.

Because of these analyses we are able to generate transgenic mice that express their transgene in almost any desired tissue by combining regulatory regions and coding sequences derived from different genes. Table 1 shows the different transgene constructs we have used over the past few years to perform tissue/cell type specific expression of heterologous genes in transgenic mice. However, the unpredictable site of integration of the transgene can affect the proper functioning of the transgene, for instance by interference with regulatory DNA elements at the integration site. The number of DNA copies incorporated can vary from one to several hundred. No relation was observed between the number of copies and the level of expression in most cases. Recently, specific DNA regions have been identified, in a few genes, the Locus Control Regions (LCR), that if present adjacent to a transgene, control transgene expression in a copy number-dependent and more or less integration site independent fashion. This LCR may allow us to produce large amounts of a specific protein in a transgenic animal (Dillon and Grosveld 1991).

The cis-acting regulatory DNA sequences of a gene are recognized by trans-acting transcription-factors, proteins that regulate the transcription of a gene. Different cell types have different sets of transcription factors that may change during the lifespan of the cell. The disruption of a gene that encodes a transcription factor will deregulate the expression of the genes that are controlled by this factor. We recently disrupted the putative transcription factor TCF-1 by homologous recombination in embryonic stem cells. The 'null mutants' show a disturbed T lymphocyte development (Verbeek *et al.* 1995).

Table 1. Tissue specific expression of transgenes

regulatory sequences	coding sequence	tissue/cell type
human CD2 promoter,	human FcγRIIA	blood/T lymphocytes
murine Ig promoter, Ig heavy chain enhancer (Eμ) 3' enhancer	human Ig heavy chain light chain	blood/B lymphocytes
murine tyrosinase	human gp100 cDNA	skin/melanocytes
murine L7 complete gene	B50 cDNA	brain/Purkinje cells
human FcγRI complete gene	human FcγRI	blood/monocytes
mouse Thy-1,2 promoter Ig heavy chain enhancer (Eμ)	human TCF-1 cDNA	blood/B- and T lymphocytes
rat Insulin II promoter	human IAPP rat IAPP	pancreas/ß-cells

An HMG-box-containing T-cell factor required for thymocyte differentiation: Cellular differentiation is mainly based on differential gene expression that, in turn, is controlled by specific transcription factors. Phenotypic changes that accompany differentiation have been described in great detail in the immune system which is one of the best defined cellular systems in higher eukaryotes. Recently, several novel transcription factors have been proposed to control cellular differentiation in the immune system, on the basis of their restricted expression patterns and their affinity for cell-lineage specific enhancers. One of them is TCF-1, a nuclear protein with a single DNA-binding HMG-box. During embryogenesis the *Tcf*-1 gene displays a complex expression pattern. By contrast, postnatally, *Tcf*-1 is uniquely expressed by mature T lymphocytes.

To study the biological functioning of the HMG-box proteins in general and that of the putative transcription factor TCF-1 in particular we disrupted the *Tcf*-1 gene by homologous recombination in mouse embryonic stem cells. In the resulting homozygous *Tcf*-1 mutant mice the thymus was the only organ that displayed gross morphological abnormalities. No numerical or phenotypical abnormalities were found in the B cell compartments of the periphery. However, at all ages, all homozygous *Tcf*-1 mutant mice showed a thymus of reduced size. Thymocyte development appears blocked at the transition of the immature $CD8^+/CD3^-/CD4^-$ single-positive (ISP) to the $CD4^+/CD8^+$ double-positive (DP) stage. Contrasting with the normal situation, ISP cells in mutant animals are predominantly not in cell cycle. Decreased numbers of immunocompetent T cells are present in peripheral lymphoid organs. We conclude that Tcf-1 controls an important step in thymocyte differentiation.

Oncology

The development of tumours is a multistep process (Berns *et al.* 1989a). Each step in tumour progression is thought to be the result of the mutation, activation or deletion of particular genes. The restricted number of genes that have the potential to play a role in tumour development can be divided in two groups: the (proto-)oncogenes and the tumour suppressor genes. Proto-oncogenes become involved in tumour development (become real oncogenes) by dominant genetic alterations resulting in deregulation of their expression and/or modification of their protein products. In contrast, tumour suppressor genes become involved in tumour development by recessive mutations. The complete absence of the normal biological activity of these genes (caused by recessive mutations in both alleles) may result in a high incidence of tumour development. In man, recessive mutations in at least 10 different loci are associated with an inherited predisposition for different tumour types.

Transgenic mice have largely contributed to the recent rapid accumulation of knowledge about cancer on the molecular level (Berns *et al.* 1989b, 1991).

Proto-oncogenes and tumour suppressor genes are studied in transgenic mice using the following approaches:

- The generation of transgenic mice that mimic the deregulated expression or the expression of mutated forms of proto-oncogenes as found in certain tumours (Van Lohuizen *et al.* 1989). A criterium for the oncogenic potential of a putative proto-oncogene *in vivo* is the ability of the activated (oncogenic) form of the gene to enhance spontaneous tumour incidence in transgenic mice. Transgenic mice are generally accepted as an *in vivo* tumorigenicity assay.
- Targeted disruption of tumour suppressor genes by homologous recombination, for instance the Rb (Clarke *et al.* 1992) and p53 (Livingstone *et*

al. 1992) gene. A restriction might be that the function of these genes is not completely the same in mice and man.
- Oncogenes that cooperate in tumour development *in vivo* can be identified in several ways (Berns *et al.* 1989b, 1991):
- The generation of transgenic mouse lines that express the activated form of two or more different proto-oncogenes (Verbeek *et al.* 1991; Moroy *et al.* 1991).
- Exposure of transgenic mice with an activated oncogene to different types of mutagenic agents like tumour viruses (Van Lohuizen *et al.* 1989, 1991), chemical compounds (Breuer *et al.* 1989, 1991) etc. By this approach several novel genes that can play a role in tumour development together with well known oncogenes were identified.

Taken together, these approaches can help us to understand in molecular terms the alterations associated both with tumour initiation and also with tumour progression.

Furthermore, transgenic mice with activated oncogenes can be used in *in vivo* carcinogenicity assays. Toxicological assays with 'normal' mice are time consuming and use many mice, because the incidence of induced tumour development is often low. However, transgenic mice with certain oncogenes (e.g. pim-1 gene) proved to be much more susceptible to the carcinogenic/mutagenic potential of compounds then non-transgenic control mice (Breuer *et al.* 1989). Our recent strategy to develop an *in vivo* short-term carcinogenicity assay is based on the assumption that mice with a defect in the DNA repair system will show an increased sensitivity for a broad spectrum of mutagenic agents. Therefore, we disrupted the XPA gene involved in DNA nucleotide excision repair by homologous recombination in embryonic stem cells. The 'null mutants' show an increased susceptibility to different genotoxic agents (De Vries *et al.* 1995).

Increased susceptibility to ultraviolet-B and carcinogens of mice lacking the DNA excision repair gene XPA: Defects in the DNA nucleotide excision repair gene *XPA* are associated with one of the most severe forms of the hereditary human disorder Xeroderma Pigmentosum. Nucleotide Excision Repair (NER) is involved in the removal of a wide spectrum of DNA lesions, induced by UV radiation and numerous genotoxic chemical agents. The XPA protein is assumed to be part of the complex that recognizes primary DNA damage. Mutations within exon 3, encoding the C4 type zinc-finger domain which is involved in binding to damaged DNA, severely diminish the biological activity of the human XPA protein *in vitro*. Moreover, mutations are found within this domain in XP-A patients. Isolated fibroblasts of these patients show a severely reduced NER activity. Due to the defective NER, the XP-A patients display a >1000-fold higher risk to develop sunlight-induced skin-cancer. An increased incidence of other type of tumours is also reported. An animal model for XP-A provides an *in vivo* system to study DNA damage repair as well as tumourigenesis resulting from defects in this system.

Therefore, we introduced a deletion spanning the exons 3 and 4 in the mouse XPA gene by homologous recombination in embryonic stem (ES) cells. Embryonic fibroblasts of the (otherwise) normal mutant mice have no detectable capacity to repair UV- and chemical-induced DNA damage *in vitro* resulting in a sensitivity to these agents comparable to that found with fibroblasts isolated from XP-A patients. In contrast to wild-type mice, the mutant mice developed skin- and eye-tumours at high frequency after low dose UV-B radiation or skin-tumours after low dose dermal exposure to the genotoxic chemical carcinogen 7,12-dimethyl benz[a]anthracene (DMBA).

From the lack of capacity to repair induced DNA damage and the highly increased sensitivity to genotoxic agents resulting in an increased incidence of tumours we conclude that our $XPA^{-/-}$ mice strongly mimic the phenotype of Xeroderma Pigmentosum patients. We therefore believe that this mouse model is useful to explain the function of the XPA protein in NER, and the role of NER in the process of spontaneous or induced carcinogenesis. More-over, our data also indicate that the XPA-deficient mice might be useful for sensitive short-term *in vivo* carcinogenicity testing of genotoxic agents. The preliminary results from continuing experiments with benzo[a]pyrene (oral administration) seem to support this.

Immunology

Higher organisms are protected against the threat of harmful foreign compounds and all different kinds of infectious agents (bacteria, viruses, fungi, protozoa and even larger parasites like worms) by their immune system. To be functional this system must be able to discriminate between what is foreign and what is a normal part of the organism to which itself belongs. In other words the immune system has to be tolerant to the tissues of the organism to which it belongs and has to destroy and remove all kinds of foreign agents. Both, self-tolerance and recognition of non-self by the immune system is based on cell to cell interactions involving receptors that either present or receive information about structural components of cells an compounds.

Particular cell types of the immune system, the B and T lymphocytes have receptors that recognize 'non-self' structures: foreign antigens. One single lymphocyte recognizes only one single foreign structure by its unique receptor. To prevent killing of own healthy cells and in the mean time to be protected against infections the organism has to develop a large repertoire of different unique T and B lymphocytes that recognizes foreign antigens and is tolerant to 'self'.

How is the B and T cell repertoire generated? How is the body provided with receptors to recognize foreign antigens that were never seen before? How is self-tolerance induced? These intriguing questions inspired immunologists during several decades to put forwards many contradictory theories. Although the basic mechanisms of the generation of the immune repertoire were already resolved, transgenic mice with unique B or T cell receptor genes provide new insights into the molecular mechanisms underlying the very complex development of the immune repertoire and the response of the immune system upon infection.

Immunology appears to be a particular fertile field for genetic analysis by gene targeting:

- the immune system is by far the best characterized cellular system in mammals;
- extensive molecular analysis has identified many of the genetic components responsible for immunological diversity (repertoire); and
- since animals with a defective immune system are still viable, analysis of many of the null phenotypes should be simplified.

Many genes involved in the immune system are knocked-out already in the mouse: for instance, the genes that encode for all chains of the T- and B cell receptor complex, most cytokines and cytokine receptors.

Transgenic mice with rearranged or mutated B- or T cell receptor genes (Kitamura *et al.* 1991; Von Boehmer and Kisielow 1990; Ishida *et al.* 1990; Bonneville *et al.* 1989, 1990; Tonegawa *et al.* 1989), with overexpressed or mutated cytokine genes (Jacob 1994) have provided new insights into the molecular mechanisms underlying the very complex development of the immune repertoire and the humoral and cellular response of the immune system to infection. The acquired knowledge will be crucial in designing new strategies to regain control over the immune system in cases where it has lost its specificity, as in autoimmune diseases, as well as to reinforce its capacity to restrain disease processes such as metastasizing cancer, chronic infectious diseases and degenerative neurological disorders.

In order to study the contribution of FcγRIII in the *in vivo* immune response, we generated mice lacking the ligand-binding chain of this receptor (Hazenbos *et al.* 1996).

FcγRIII deficient mice show impaired immune functions: Receptors for the Fc portion of IgG (FcγR) are expressed on a variety of haematopoeitic cells. In mice, three distinct FcγR can be recognized: FcγRI, FcγRII and FcγRIII. The precise role of each of these FcγR in immune reactions has not been clarified due to the lack of an experimental model which enables a clear functional distinction between different FcγR.

Recently a knock-out (KO) mouse lacking the common accessory γ-chain of FcγRI, FcγRIII, and FcεR has been described, which exhibits pleiotropic defects in immune functions. In order to study the contribution of FcγRIII in the *in vivo* immune response more specifically, KO mice lacking the ligand-binding chain of this receptor were generated by homologous recombination in embryonic stem cells.

Macrophages from peritoneal exudate of FcγRIII-KO mice exhibited a marked defect in the phagocytosis of IgG-opsonized erythrocytes. When triggered with an anti-FcγRII/III antibody, degranulation of mast cells from FcγRIII-KO mice was negligible as compared with wild-type mast cells. Both phagocytosis and mast cell degranulation mediated by FcγRI, or by FcεR, respectively, was intact in KO mice. Antibody-dependent-cell-mediated-cytotoxicity (ADCC) by NK cells from KO mice was strongly impaired. These results demonstrate that FcγRIII is a crucial receptor involved in various IgG-mediated functions of the immune system, and that these KO mice provide a useful tool to study the function of FcγRIII in such processes *in vivo*.

Antigen targeting to myeloid-specific human FcγRI/CD64 triggers enhanced antibody responses in transgenic mice: Antigen presentation by human monocytes was recently found to be enhanced *in vitro* through the high-affinity Fc receptor for IgG FcγRI which is exclusively present on myeloid cells. To evaluate a comparable role of FcγRI in antigen presentation *in vivo*, we generated human FcγRI transgenic mice (Heijnen *et al.* 1996). Under the control of its own regulatory sequences, human FcγRI was expressed selectively on myeloid cells at physiological expression levels. As in man, expression was properly regulated by the cytokines IFN-γ, G-CSF, Il-4 and Il-10, and was up-regulated during inflammation. The human receptor expressed by the murine macrophages bound monomeric IgG and mediated particle phagocytosis and IgG complex internalization. To evaluate whether specific targeting of antigen to FcγRI can induce enhanced antibody responses, mice were immunized with an anti-human FcγRI antibody containing antigenic determinants. Transgenic mice produced antigen-specific antibody responses with high IgG1 titres, and substantial IgG2a and IgG2b responses. These data demonstrate that human FcγRI on myeloid cells is highly active in mediating enhanced antigen presentation *in vivo*, and show that anti-FcγRI monoclonal antibodies are promising vaccine adjuvants. Moreover, this mouse model can be used to evaluate in an animal model bi-specific tumour therapy strategies. In this approach the tumour-killing potency of neutrophils is activated by cross-linking FcγRI on these cells with a specific surface antigen of the tumour.

Disease Models
The mouse shares with man most metabolic pathways and physiological processes, making the mouse a valuable model system to study human diseases. Good animal models of human diseases are powerful tools in medical research. However, often no useful animal model exists. The technology of transgenesis provides us with the opportunities to generate the desired human disease models. There are different approaches:

- Indirect: the introduction of a transgene or a mutation that creates physiological conditions that mimic the abnormal physiological conditions in certain human diseases. This type of models allow us to study the long term effects of the abnormalities and to develop and test therapeutics that combat the symptoms.
- Direct: the deregulated expression of a gene that is somehow associated with the initiation or/and maintenance of a human disease. To mimic a mutation, by using targeted mutagenesis, in the mouse genome that in humans has been identified as to be responsible for a certain disease.

There is a major thrust to use targeted mutagenesis to generate mouse models for human diseases for which the responsible mutations have been identified and mapped, such as Xeroderma Pigmentosum (De Vries *et al.* 1995), cystic fibrosis (Snouwaert *et al.* 1992), and others. The expectations are high, but only the experiment will teach us whether the specific mutant mice can serve as valuable model systems for the corresponding diseases in man and subsequently for the development of therapies (Hyde *et al.* 1993). In transgenic mouse models of some other more complex diseases like non-insulin dependent diabetes (NIDDM) we showed that the involvement of a distinct gene can not easily be assessed by over-expressing that gene (Höppener *et al.* 1993). But this transgenic model will still contribute to the understanding of at least some aspects of the complex disease (De Koning *et al.* 1994).

Intra- and extracellular amyloid fibrils are formed in cultured pancreatic islets of transgenic mice expressing human islet amyloid polypeptide: Non-insulin-dependent diabetes (NIDDM) is characterised by hyperglycaemia, impaired insulin secretion and deposition of amyloid in pancreatic islets. Islet amyloid accumulates between endocrine cells and islet capillaries. Progressive deposition of amyloid fibrils contributes to decreased islet function and loss of insulin-secreting ß cells in the course of the disease. Islet amyloid polypeptide (IAPP, amylin) is the major peptide component of islet amyloid. IAPP is synthesised in pancreatic ß-cells and co-secreted with insulin. The factors causing polymerisation of IAPP to form fibrils are unknown. Species-specific variations in the amino-acid sequence of IAPP determines the capacity of the molecule to form fibrils: human IAPP (hI-APP) is amyloido-genic whereas rat (rIAPP) and the identical mouse IAPP are not.

To examine the causative factors for amyloidogenesis, transgenic mice were generated to express the genes for either the human or the rat IAPP. Fibrillar islet amyloid was not detectable by light or electron microscopy in these normoglycaemic, normoinsulinaemic mice, despite high circulating concentrations of IAPP. However, when islets from hIAPP mice were cultured *in vitro*, in the presence of 11.1, 16.7 and 28 mM glucose, amyloid fibrils formed within two days. Amyloid fibrils were not formed in islets from rIAPP mice or non-transgenic controls or at 4.0 mM glucose. When cultured in 11.1 mM glucose IAPP secretion from 'hIAPP islets' was elevated compared to that from 'non-transgenic islets', whereas insulin secretion was similar in both. IAPP immunoreactive fibrils were present between cells and in deep invaginations of ß cells suggesting that fibril formation occurs during exocytosis of ß cell granules. Despite the large concentrations of fibrils no evidence for ß cell destruction was observed within 21 days. This *in vitro* model will permit determination of factors involved in fibrillogenesis and in the progressive accumulation of amyloid aggregation.

Ethical Aspects

This overview of part of the research with transgenic mice generated in only one transgenic mouse facility already demonstrates the strong power of the transgenic mouse model as a scientific tool. Nevertheless, scientists cannot ignore the increasing concern of the general public with respect to genetic manipulation. Many of the fears can be attributed to a lack of knowledge on this matter. In addition, there is, quite understandably, concern about misuse and exploitation of animals. Although clear guidelines are required to settle both practical as well as ethical issues with respect to germ-line manipulation of animals in general, there is a clear difference between the scientific applications of the technology and the generation of transgenic livestock. I would like to discuss a few aspects.

The Lack of Control of the Integration Site of the Transgene

Most of the gene transfer methods used today cannot target transgenes efficiently to a specific chromosomal locus. So there is a risk of genetic damage. Therefore scientists will always generate several independent lines of transgenic mice that differ only in the integration site of the transgene. If the different transgenic lines show the same phenotype, non-specific effects caused by the integration of the transgene can be excluded. A transgenic line that shows an additional change in phenotype will not be bred further.

However, when transgenesis is considered for the generation of new herds of livestock, special attention has to be devoted to the "genetic fitness" of the transgenic founders (apart from the obvious requirement to maintain sufficient genetic variation in the various species of domestic animals, a point which is also of importance in classical breeding programs). Classical breeding is based on selection of external characteristics. A disadvantage of this method is that together with the desired characteristics other undesired hidden genetically linked characteristics are selected. Because of the restricted knowledge about the genome of livestock we cannot identify these undesired genes. Sometimes these genes strongly influence the welfare of the animal. By using the technology of transgenesis only the well defined gene that encodes the desired characteristic will be introduced in the genome. Moreover, the integration site of a transgene in the genome can easily be identified by cloning the flanking sequence using the transgene as a tag. Integration site analysis allows the selection of transgenic animals with undamaged endogenous genes.

In developmental biology, integration mutagenesis by transgenes is even used as a method to identify novel genes involved in embryonic development. Transgenic mouse lines in which the transgene disrupts a gene essential for embryonic development will never give rise to homozygous offspring. In the mouse, the site of integration can be controlled by using the technology of gene transfer into ES cells that can give rise to a high frequency of homologous recombination. In livestock, the real breakthrough in transgenesis might come from the generation of ES cell lines from a variety of mammals, allowing application of the already developed targeting strategies in mice.

The advantage of the well defined genetic alterations induced by the technique of homologous recombination is clearly demonstrated by the replacement of the widely used SCID mouse model by the Rag1/2 knockout mouse model in present immunological studies. Both mouse models show about the same phenotype of immune deficiency. However, in the 'natural' occurring SCID mouse it cannot be excluded that unidentified additional mutations exist.

Heterologous Transgenes and the Identity of the Transgenic Animal

The scientific transgenic mouse models generated with heterologous genes are mainly mouse models with human genes for medical model studies. The efficacy of pharmaceuticals is based on their interaction with the right human target molecule. In the past, the clinical trial, with its strong ethical constraints, was the only way to test efficacy and side effects *in vivo*. Now, pharmaceutical tests in a transgenic mouse model that expresses the human target molecule in the right site in the body can precede and partly replace the clinical trial. The transgenic mouse model (described previously) that expresses the human FcγRI is an example of this approach. The international community faces the pharmaceutical industry with a complex problem. On one hand, the increasing demands for detailed analysis of the biological activity and the side effects of new pharmaceuticals requires extensive *in vivo* research in well defined animal model systems. On the other hand, the wish to reduce the use of laboratory animals and to reduce the generation of transgenic animal models hampers the development of these *in vivo* tests.

In general, the heterologous genes that are considered to become transgenes in livestock are strongly conserved in evolution. In a single transgenic experiment, the genetic alteration is very restricted: a single gene that is homologous for more than 70% with an endogenous gene, is added to the genome that contains more than 50,000 genes. It cannot be expected that this will affect the identity of the animal.

The Effects of the Expression of the Transgene

So far there is no proof that the technology of transgenesis as such influences the well-being of the transgenic animal. The main goal of transgenic models is to generate an animal model with a detectably-altered phenotype. This phenotype may influence the welfare of the transgenic animal. However, with respect to animal welfare, *in vivo* models generated by using the technique of transgenesis do not differ from *in vivo* models generated by using mutagenic agents and the selection procedures of classical breeding. Therefore the decision to permit the generation of animal models resulting in suffering of these animals has to be based not on the technology used to generate the model, but only on the expected phenotype of the model.

The main goal of transgenic livestock is to generate an animal as a bioreactor for pharmaceutical proteins. It might be important for the well-being of the animal that the transgene product does not interfere with physiological processes of the animal. Specific targeting to an organ compartment like the udder or the bladder can prevent such interference.

Risk to the Environment

Transgenic mammals will, in general, pose no risk to the environment. Also the chance that the transgene is inadvertently introduced into the "wild" population is extremely small: for transgenic mice such spreading will be self-limiting due to the inability of laboratory mice to survive outside a highly protected environment; for larger domestic animals physical containment can be sufficiently assured to prevent undesired genetic mixing. Obvious exceptions are transgenic animals carrying infectious agents as transgenes. Mice carrying the intact HIV genome as transgene might serve as an example. These animals require a special barrier facility.

Acknowledgements

I thank my collaborators, the groups of Dr. H.C. Clevers (TCF-1), Dr. H. van Steeg (XPA), Dr. Anne Clark & Dr. J. Höppener (IAPP), and Dr. J.G.J. van de Winkel (hFcγRI) for their scientific contributions. I thank F.M.A Hofhuis for his excellent technical support (generation of the mice) and Dr. A. Snijders for critical reading the manuscript.

References

Berns A, Breuer M, Verbeek S, Van Lohuizen M (1989a) Synergism between oncogenes in T-cell lymphomagenesis. In: NATO ASI Series, Vol. 34, Vectors as Tools for the Study of Normal and Abnormal Growth and Differentiation (Lother H et al. eds) Berlin-Heidelberg: Springer Verlag, pp. 344-353.

Berns A, Breuer M, Verbeek S, Van Lohuizen M (1989b) Transgenic mice as a means to study synergism between oncogenes. International Journal of Cancer, supplement 4, 22-25.

Berns A, Van Lohuizen M, Verbeek S, Domen J, Saris C (1991) Transgenic mice as a model system to study synergism between oncogenes. Origins of Human Cancer: A Comprehensive Review, Cold Spring Harbor (NY): Cold Spring Harbor Laboratory Press.

Bonneville M, Ishida I, Mombaerts P, Katsuki M, Verbeek S, Berns A, Tonegawa S (1989) Blockade of αß T cell development by τδ TCR transgenes. Nature 342, 931-934.

Bonneville M, Ishida, I, Itohara, S, Verbeek S, Berns A, Kanagawa O, Haas W, Tonegawa S (1990) Self-tolerance to transgenic τδ T cells by intrathymic inactivation. Nature 344, 163-165.

Breuer M, Slebos R, Verbeek JS, Van Lohuizen M, Wientjes E, Berns A (1989) Very high frequency of lymphoma induction by a chemical carcinogen in pim-1 transgenic mice. Nature 340, 61-63.

Breuer M, Slebos R, Wientjens E, Verbeek S, Berns A (1991) Carcinogen induced lymphomagenesis in pim-1 transgenic mice: dose dependence and involvement of myc and ras. Cancer Research, 958-963.

Capecchi M R (1989). The new mouse genetics: Altering the genome by gene targeting. Trends Genet. 5,70-76.

Clarke AR, Robanus-Maandag E, Van Roon M, Van der Lught NMT, Van der Valk M, Hooper ML, Berns A, Te Riele H (1992). Requirement for a functional Rb-1 gene in murine development. Nature 359, 328-330.

De Koning EJP, Morris ER, Hofhuis FMA, Posthuma G, Höppener JWM, Morris JF, Capel PJA, Clark A, Verbeek JS (1994) Intra- and extracellular amyloid fibrils are formed in cultured pancreatic islets of transgenic mice expressing human islet amyloid polypeptide. Proceedings of the National Academy of Sciences (USA) 91, 8467-8471.

De Vries A, Van Oostrom CTM, Hofhuis FMA, Dortant PM, Berg RJW, De Gruijl FR, Wester PW, Van Kreijl CF, Capel JA, Van Steeg H, Verbeek SJ (1995) Increased susceptibility to ultraviolet-B and carcinogens of mice lacking the DNA excision repair gene XPA. Nature 377, 169-173.

Dillon N, and Grosveld F (1991) Human gamma-globin genes silenced independently of other genes in the ß-globin locus. Nature 350, 252-254.

Hanahan D (1989) Transgenic mice as probes into complex systems. Science 246, 1265-1275.

Hazenbos W, Gessner E, Hofhuis F, Kuijpers H, Heijnen I, Capel P, Van de Winkel J., Daëron M, Verbeek S (1996) FcγRIII deficient mice show impaired immunological functions. Submitted.

Heijnen I, Van Vugt M, Fanger NA, Graziano RF, De Wit TPM, Hofhuis FMA, Guyre,PM, Capel PJA, Verbeek JS, Van de Winkel JGJ (1996) Antigen targeting to myeloid-specific human FcγRI/CD64 triggers enhanced antibody responses in transgenic mice. Journal of Clinical Investigation 97, 331-338.

Höppener JWM, Verbeek JS, De Koning EJP, Oosterwijk C, Van Hulst KL, Visser-Vernooy HJ, Hofhuis FMA, Van Gaalen S, Berends MJH, Hackeng WHL, Jansz HS, Morris JF, Clark A, Capel PJA, Lips CJM (1993) Chronic overproduction of islet amyloid polypeptide/amylin in transgenic mice: lysosomal localisation of human islet amyloid polypeptide and lack of marked hyperglycaemia or hyperinsulinaemia. Diabetalogia 36, 1258-1265.

Hyde SC, Gill DR, Higgins CF, Trezise AEO, MacVinish LJ, Cuthbert AW, Ratcliff R, Evans MJ, Colledge WH (1993) Correction of the ion transport defect in cystic fibrosis transgenic mice by gene therapy. Nature 362, 250-255.

Ishida I, Verbeek S, Bonneville M, Itohara S, Berns A, Tonegawa S (1990) T-cell receptor γδ and γ transgenic mice suggest a role of a γ gene silencer in the generation of αß T cells. Proceedings of the National Academy of Science (USA) 87, 3067-3071.

Jacob CO (ed) (1994) Overexpression and knockout of cytokines in transgenic mice. San Diego: Academic Press Inc.

Jaenisch R (1988) Transgenic animals. Science 240, 1468-1474.

Joyner AL (ed) (1993) Gene Targeting, A Practical Approach. New York: Oxford University Press Inc.

Kitamura D, Roes J, Kühn R, Rajewsky K (1991) A B cell-deficient mouse by targeted disruption of the membrane exon of the immunoglobulin μ chain gene. Nature 350, 423-426.

Livingstone LR, White A, Spouse J, Livanos E, Jacks T Tisty TD (1992) Altered cell cycle arrest and gene amplification potential accompany loss of wild-type p53. Cell 70, 923-935.

Moroy T, Verbeek JS, Zornig M, Ma A, Achacoso P, Berns A Alt FW (1991) EμN- and EμL-myc cooperate with Eμ pim-1 to generate lymphoid tumors at high frequency in double transgenic mice. Oncogene 6, 1941-1948.

Palmiter RD, Brinster RL (1986) Germline transformation of mice. Ann. Rev. Genet. 20, 465-499.

Snouwaert JN, Brigman KK, Latour AM, Malouf NN, Boucher RC, Smithies O, Koller, BH (1992) An animal model for Cystic Fibrosis made by gene targeting. Science 257, 1083-1088.

Tonegawa S, Berns A, Bonneville M, Farr AG, Ishida I, Ito K, Itohara S, Janeway CA, Kanagawa O, Katsuki M, Kubo R, Lafaille JJ, Mombaerts P, Murphy DB, Nakanishi N, Takagaki Y, Van Kaer L, Verbeek S (1989) Diversity, Development, Ligands, and Probable Functions of τδ T cells. Cold Spring Harbour Symposium Quant. Biology, volume LIV. Cold Spring Harbor (NY): Cold Spring Harbor Laboratory Press.

Van Lohuizen M, Verbeek JS, Krimpenfort P, Radaszkiewicz T, Berns A (1989) Predisposition to lymphomagenesis in pim-1 transgenic mice: cooperation with c-myc and N-myc in MuLV induced tumors. Cell 56, 673-682.

Van Lohuizen M, Verbeek S, Scheijen B, Wientjens E, Van der Gulden H, and Berns A (1991) Identification of cooperating oncogenes in Eμ-myc transgenic mice by provirus tagging. Cell 65, 737-752.

Verbeek S, Van Lohuizen M, Van der Valk M, Domen J, Kraal G, Berns (1991) A Mice bearing the Eμ-myc and Eμ-pim-1 transgenes develop pre-B cell leukemia prenatally. Molecular Cell Biology 11, 1176-1179.

Verbeek JS, Izon D, Hofhuis FMA, Robanus-Maandag E, Te Riele H, Van de Wetering M, Oosterwegel M, Wilson A, MacDonald HR, Clevers H (1995) An HMG-box- containing T-cell factor required for thymocyte differentiation. Nature 374, 70-74.

Von Boehmer H, Kisielow P (1990) Self-nonself discrimination by T cells. Science 248, 1369-1373.

Introduction of Defined Mutations into the Mouse Germline

W. Müller

Summary

Murine embryonic stem cell lines can be established from the inner cell mass of murine blastocysts. They can be kept undifferentiated in long term culture and are amenable to genetic alterations. Embryonic stem cells can be reintroduced into mouse blastocysts and will participate in the development of the mouse, including the mouse germline. Using various selection procedures, rare genetic events, such as planned homologous recombinations at defined sites, can be selected and mouse strains with precise mutations can be generated. By introducing genetic elements such as specific recombination target sequences to the target gene and by utilizing the corresponding recombinases *in vitro* and *in vivo*, cell type specific gene alterations can be achieved. This new technology provides tools to control for animal welfare on the genetic level.

Introduction

The mouse as a model organism has given insights into the development and function of the vertebrates in various areas of basic research such as developmental biology, cancer research, neurobiology and immunology. These findings, which can be in part extrapolated to humans, will therefore have impact on modern medicine. Here, I want to describe approaches to genetically modify the mouse genome in a predefined way and to discuss welfare aspects which must be taken into consideration when working with mouse mutants.

The Embryonic Stem (ES) Cell Technology

Derived from the inner cell mass of mouse blastocysts, embryonic stem (ES) cell lines have been established from various mouse strains. These lines can be cultured for an extended period of time *in vitro* on a layer of embryonic fibroblasts in the presence of a cytokine, leucocyte inhibitory factor. In this culture system, cell differentiation is prevented and the ES cells maintain their pluripotent capacity. This means that, when implanted into mouse blastocysts, ES cells are able to participate in the development of the mouse and to differentiate into all tissues of the developing mouse. When introduced into blastocysts of a mouse strain different from the strain from which the ES cell line was established, it is possible to distinguish between cells which are ES cell-derived and cells which are blastocyst-derived. ES cells will also participate in the development of germ cells and by this, a mouse line can be established which carries the genetic information of the ES cell line. This opens up the possibility of altering the genome of the mouse by introducing new DNA into ES cells, to select mutant ES-lines and to subsequently establish mouse mutant lines. The ES cell technology is described in detail in various books (Hooper 1992, Joyner 1993, Robertson 1987).

Generation of Transgenic Mouse Lines Using ES Cell Technology

ES cells can be transfected with DNA and this DNA will integrate randomly into the ES cell genome. Once integrated the new DNA (transgene) will remain stable within a particular ES clone and can be introduced into the mouse germ line by introducing the ES clone into mouse blastocysts and breeding of the chimaeric mice. The ES cell technology can be regarded as an alternative to microinjection of DNA into mouse oocytes for generating transgenic lines. The use of ES cells has several advantages compared to the microinjection procedure. When microinjection is used, every transgenic mouse born represents the founder of a new strain and the effect of the transgene can only be analyzed in its first offspring. With the ES cell approach, even before the generation of the chimaeric mouse, the ES cells can be analyzed extensively by molecular biological methods. For example, a series of ES cell clones can be selected with various numbers of trans gene copies. Next, transgene effects can be studied in cell culture and in the chimaeric mouse. If the phenotype is as expected, the transgene can be introduced into the mouse germline by breeding. The modification (a particular transgene insertion point and a particular copy number) is preserved in the ES cell clone and new chimaeric mice can be generated when required. The ES cell clones can be frozen in liquid nitrogen and remain available practically for an unlimited time. In contrast, with pronucleus injection, the mouse line has to be cryo-preserved as embryos.

Using the ES cell technique, several integration sites of a particular construct can be tested and only 1 or 2 of many ES-cell clones are later chosen for the establishment of a mouse line. If one wants to study regulatory DNA-elements for directing expression of a transgene, these elements can be analyzed in the chimaeric mouse and no mouse lines have to be established. In the ES cell system, cells of the blastocysts can compensate for possible detrimental effects of the transgene present in the ES cell in the chimaeric mouse. In the past, ES cells have been used to successfully generate transgenic mouse lines (Gossler *et al.* 1986, Müller *et al.* 1991, Stewart *et al.* 1987).

Introduction of Targeted Mutations into ES Cells

DNA sequences, which are in part homologous to the genome of the ES cells, can integrate in two ways into the ES-cell genome. Most frequently, the DNA will insert at a random place in the ES-cell genome (random integration). At a somewhat lower frequency, the DNA will recombine with its homologous sequence within the ES-cell (homologous recombination). By constructing suitable DNA-sequences and by selecting for the homologous recombination event in the transfected ES-cell, various types of mutations can be introduced to the ES-cell genome (and through this into the mouse genome) (Capecchi 1989). It is possible to inactivate genes, to replace genes, to repair genes, to alter gene regulatory sequences, to introduce sequences with particular properties (promoters, target sites for recombination) and so forth.

By using standard techniques, the resistant marker genes, used to select for ES-cells which carry the new genetic information, remain at the site of homologous recombination, imposing restrictions on the types of alterations which can be introduced at a given site. A second limitation of this procedure is that one might introduce a mutation which is detrimental for mouse development and the result of the genetic alteration cannot be studied. Both limitations can be overcome by the use of the *Cre/loxP* technology.

Introduction of Defined Mutations into ES-cells by the Cre/loxP Technology
The *Cre/loxP* technology is based on the bacteriophage P1 recombinase, *Cre*, and its target sequence, *loxP*, a 34 base pair sequence motive present

in the P1-phage genome. *Cre* recognizes its target sequence and will perform recombination when a second *loxP* site is found. The outcome of the recombination event will depend on the orientation of the *loxP* sites. *Cre* can mediate DNA-deletion, inversion or insertion. If mutant *loxP* sequences are used this process can be made irreversible. When the wild type *loxP* sequence is used, the recombination process can be re-versible.

It is important to note that the bacteriophage *Cre* recombinase can act without cofactors, functions in eucaryotic cells, works also as a gene fusion product and its target sequence (*loxP*) is apparently not present in the eucaryont genome. *Cre/loxP* allows now various modifications to the original protocol of introducing defined mutations into the ES-cells by homologous recombination. By flanking the selection marker gene with *loxP* sites, the selection marker can be removed after the identification of the homologous recombination event by transient expression of *Cre* in the ES cell line. By introducing additional *loxP* sites into the genome, small or even large deletions of DNA can be made on a particular chromosome and by the design of proper targeting vectors, gene segments can be replaced allowing the introduction of small mutations (point mutations, base deletion or insertion) or gene segment replacement (for example the exchange of a murine gene *exon* by its human counterpart (Zou *et al.* 1994)).

Somatic Gene Alteration Using the Cre/loxP Technology

The bacteriophage *Cre* recombinase can act in eucaryotic cells. In the previous section, gene alterations using *Cre/loxP* were described utilizing *Cre* in the ES-cell line. *Cre* can also be expressed and function in the mouse as a transgene. By homologous recombination, gene segments can be flanked by *loxP* sites and these modified genes can be introduced into the mouse germline. If the sites of the loxP sites are carefully chosen, the modified gene segment will still function as the original gene segment and a mouse mutant carrying this gene segment will have no phenotype. When this mouse mutant is crossed to a *Cre*-transgenic mouse line, then in all cells, in which the recombinase *Cre* is expressed, the *loxP* flanked gene segment will be deleted, in the other cells the gene segment will be unaltered and the gene remains functional (Gu *et al.* 1994). By this procedure, the function of a gene can be studied in a particular cell type.

The regulation of the *Cre*-transgene can also be made inducible. For example, if the recombinase *Cre* is expressed under the control of an inducible promoter, a mouse carrying the *loxP* flanked gene segment and this transgene will remain normal until the Cre-transgene expression is turned on by an inducer. After the induction of the Cre-transgene expression, the *loxP* flanked gene segment is deleted and the gene function may be abolished. This procedure allows the analysis of gene function at a particular time point (Kühn *et al.* 1995). It is also possible to combine cell type specific expression of *Cre* with inducibility. Finally, the *Cre* recombinase does not need to be a transgene. It can also be introduced into the cells by viruses or as naked DNA. By this it will be possible to restrict *Cre*-mediated deletion to a particular site. Using any of the methods described here allows the analysis of gene function in the adult mouse for genes which are essential for mouse development.

Welfare Aspects of Mouse Mutants Carrying Defined Mutations

In the past, mouse breeders have selected natural mouse mutants by searching for mice which show a phenotype. Many mutants have been selected by this way (from coat colour mutants to mice with severe deficiencies in the central nervous system or in the haematopoietic system). Some of the natural mutants have been characterized at the molecular level, others are not yet known. The new technology of generating mouse mutants carrying defined mutations is not dependent on a phenotype but dependent on the introduction of a defined mutation into a particular gene. This procedure does not automatically generate a mouse mutant with a "severe" phenotype. In fact, many mutants with defined mutations do not show an obvious phenotype and are only recognizable by molecular analysis. The reason for this is that many gene systems appear somewhat redundant and other similar genes can take over the function of the modified gene. Also the environment, in which laboratory animals are developing, is not as selective as the environment outside of the laboratory and it may be that many genes which are very important for the survival in the wild are dispensable in the laboratory. Therefore, when discussing welfare aspects of mouse mutants carrying defined mutations, one has to take into account the environment in which the animals are kept. Many mutants will not suffer under the condition of a normal mouse breeding environment and natural mutants, which carry a corresponding mutation, would live there without being noticed by the mouse breeder. On the other hand, by introducing defined mutations into the mouse germline, mice can develop which have difficulties dealing with the environment of a normal laboratory.

However, it is impossible to distinguish between welfare aspects of naturally occurring mouse mutants and mutants carrying defined mutations. As an example, I would like to discuss the welfare aspect of the natural mutant SCID (Bosma *et al.* 1983) and the Rag-1 (Mombaerts *et al.* 1992) and

Rag-2 (Shinkai *et al.* 1992) knockout mice. Both type of mouse mutants lack functional T and B lymphocytes and their immune system is not functional. Consequently, both mutants are used to receive transplants of human tissues. In a conventional animal house, both mutants would suffer and die due to infections. When kept in isolators and in a clean, not necessarily germ-free environment, the mice develop nicely and breed well. However, when both mutants are kept as heterozygous animals, they will develop normally even in a conventional animal house.

Table 1. Assessment of the welfare of mouse mutants deficient for genes expressed in the haematopoietic system (survey: December 1994)+

	Phenotype			
	[lethal]*	[severe]	[mild]	[none]
natural mutant	1	6	1	0
knockout mutant	5	36	24	14

\+ The biological consequences of molecular characterised mutations in the mouse genome of genes, which are expressed in the haematopoietic system, were grouped into four arbitrary categories. These categories assess whether the mutants will be impaired in a conventional mouse facility. The numbers represent number of mouse mutants.
The complete list will be published in the Handbook of Experimental Immunology.

* [lethal] means that the mice homozygous for the mutation die before birth. In case of a [severe] phenotype a major lymphocyte compartment of the immune system is absent or strongly impaired.
[mild] indicates that all major lymphocyte subsets are present but some of them are functionally impaired.
[none] means that the immune system is functional. The welfare of many mouse mutants of the groups [severe] and [mild] might not be impaired when kept in a specific pathogen-free environment.

In December 1994, I compiled a list of mouse mutants with alterations of genes expressed in the haematopoietic system. The list contained both natural and induced mutants. The mouse mutants were grouped into the four categories, depending on the phenotype (Table 1). As expected, all natural mutants show a phenotype, otherwise they would not have been detected. On the other hand, mutants carrying intentional introduced mutations do not necessarily show a phenotype. In fact, the number of mutants showing no phenotype is an underestimate as only mutants which were published at the time are included. Mutants, which do not show any phenotype are hard to publish and therefore were not available for the analysis. Furthermore, in the beginning of gene targeting, the choice of genes was highly selective and mutations which will result in a phenotype were favoured by the researchers.

Conclusion

The ES cell technology allows the generation of transgenic mouse lines in a much more predictable way than the standard microinjection technique. Modern technology has made it possible to introduce defined mutations into the mouse germline. This technology has proved successful for the assignment of gene functions *in vivo* in biological systems which can not be studied *in vitro*, for example the immune system. The *Cre/loxP* technology allows the analysis of gene function restricted to particular cell types and/or for particular time points. As a result, mouse mutants would potentially suffer only under particular experimental conditions and the untreated mutant strains are not impaired. Finally, the welfare aspect of mouse mutants does not depend on the way the mutation was introduced into the mouse genome (natural or induced) but depends on the phenotype caused by the mutation. Transgenesis by means of introducing defined mutations into mouse ES cells does not automatically lead to a welfare problem.

References

Bosma GC, Custer RP, Bosma MJ (1983) A severe combined immunodeficiency mutation in the mouse. Nature 301, 527-530.

Capecchi MR (1989) Altering the genome by homologous recombination. Science 244, 1288-1292.

Gossler A, Doetschman T, Korn R, Serfling E, Kemler R (1986) Transgenesis by means of blastocyst-derived embryonic stem cell lines. Proceedings of the National Academy of Sciences (USA) 83, 9065-9069.

Gu H, Marth JD, Orban PC, Mossman H, Rajewsky K (1994) Deletion of the DNA polymerase ß gene in T cells using tissue-specific gene targeting. Science 265, 103-106.

Hooper ML (1992) Embryonal stem cells. Chur: Harwood Academic Publishers.

Joyner AL (1993) Gene Targeting. New York, Tokyo: Oxford University Press.

Kühn R, Schwenk F, Aguet M, Rajewsky K (1995) Inducible gene targeting in mice. Science 269, 1427-1429.

Mombaerts P, Iacomini J, Johnson RS, Herrup K, Tonegawa S, Papaioannou VE (1992) RAG-1-deficient mice have no mature B and T lymphocytes. Cell 68, 869-877.

Müller W, Kühn R, Rajewsky K (1991) Major histocompatibility complex class II hyperexpression on B cells in interleukin 4-transgenic mice does not lead to B cell proliferation and hypergammaglobulinemia. European Journal of Immunology 21, 921-925.

Robertson EJ (1987) Teratocarcinomas and embryonic stem cells: a practical approach. Oxford, New York, Tokyo: ILR Press Limited.

Shinkai Y, Rathbun G, Lam KP, Oltz EM, Stewart V, Mendelsohn M, Charron J, Datta M, Young F, *et al.* (1992) RAG-2-deficient mice lack mature lymphocytes owing to inability to initiate V(D)J rearrangement. Cell 68, 855-867.

Stewart CL, Schuetze S, Vanek M, Wagner EF (1987) Expression of retroviral vectors in transgenic mice obtained by embryo infection. Embo Journal 7, 383-388.

Zou Y, Müller W, Gu H, Rajewsky K (1994) Cre-loxP-mediated gene replacement: a mouse strain producing humanized antibodies. Current Biology 4, 1099-1103.

Weir DM (ed) Compilation of mouse mutants with alteration of genes expressed in the haematopoietic system. In: Handbook of Experimental Immunology, 5th edition. Blackwell Scientific Publications, in press.

Growth Hormone Overproduction in Transgenic Mice: Phenotypic Alterations and Deduced Animal Models

E. Wolf and R. Wanke

Summary

Transgenic mice play a key role in biomedical research since they can be designed to be optimum animal models for studying specific scientific problems. In the early 1980s, transfer of growth hormone (GH) fusion genes to the germ-line of mice demonstrated that foreign genetic information may be stably integrated and expressed to yield a biologically active product, as evidenced by markedly stimulated growth of these mice. Consequently, GH structural genes have been used as welcome reporters for functional analyses of regulatory DNA sequences. Phenotypic characterisation of mice expressing various GH fusion genes revealed a typical spectrum of alterations. These alterations included an increased growth of body, skeleton, muscles, and internal organs; endocrine and metabolic changes; as well as pathological alterations primarily of liver and kidneys. From these alterations, which occur reproducibly in an age-dependent manner, animal models have been developed for studying the long-term effects of elevated GH, the pathogenesis of glomerulosclerosis and the mechanisms involved in hepatocarcinogenesis. In this review, we compare GH transgenic mice with the classical animal models which have been used for studying these problems with respect to both the scientific quality of these models and the animal welfare aspect. The well-being of GH transgenic mice may - at least if young animals are investigated - be less disturbed than that of the comparable animal models produced conventionally. The scientific questions currently investigated using GH transgenic mice appear to be sufficiently important for mankind to be justified on ethical criteria.

Introduction

The advent of transgenic technology led to a plethora of transgenic animal models specifically designed for studying mechanisms of gene regulation, development, pathogenesis, and the treatment of disease. Strategies for introducing genetic information into animal genomes include: pronuclear DNA-microinjection; generation of germ-line chimaeric mice using embryonic stem cells following targeted mutagenesis; as well as the use of retroviral vectors (for review, see Brem 1993). The impact of gene transfer became evident with the generation of giant transgenic mice due to the integration and expression of rat (r) or human (h) GH sequences (Palmiter *et al.* 1982; Palmiter *et al.* 1983). On the other hand, pronuclear microinjection of hGH sequences also demonstrated, for the first time, that random insertion of heterologous (= foreign) DNA may disrupt essential endogenous sequences resulting in a lethal insertional mutation (Wagner *et al.* 1983).

Subsequently, sequences coding for the growth hormones of various species have frequently been used as reporter genes for the functional analysis of regulatory DNA sequences (e.g. Ornitz *et al.* 1985; Pinkert *et al.* 1987; Zambrowicz *et al.* 1993) or for cell differentiation studies (e.g. Bry *et al.* 1994), since the expression of such transgenes can be easily detected in the level of RNA (Ehrlein *et al.* 1994) and protein (Wolf and Brem 1991). Mutated GH coding sequences have been expressed in transgenic mice to determine the function of different domains of the human and bovine (b) GH molecules *in vivo* (Chen *et al.* 1994; Chen *et al.* 1995).

Within the last decade, a number of transgenic lines expressing foreign GH sequences under the transcriptional control of various regulatory elements with different tissue specificities were created. Analyses of the phenotypes of these mice revealed a typical spectrum of alterations which are associated with GH overproduction: markedly increased body growth; visceromegaly; and disproportionate skeletal gigantism; as well as characteristic hepatic and renal alterations, the latter being the primary cause of the shortened life-span of these mice.

Therefore, GH transgenic mice are suitable models for studying the long-term effects of elevated levels of heterologous growth hormones without immunological complications. They provide insights into the physiology and pathophysiology of growth regulation. As GH transgenic mice consistently develop renal alterations, they have been accepted as a useful model for studying the pathogenesis of glomerulosclerosis. A high incidence of neoplastic alterations of the liver in GH transgenic mice also makes them a non-genotoxic model of hepatocarcinogenesis.

In this review, we give a survey of phenotypic alterations observed in GH transgenic mice and evaluate their meaning for biomedical research by comparing them with conventional animal models.

Integration and Expression of Growth Hormone Transgenes

Gene Constructs
Two types of promoter/regulatory sequences have been used to control the expression of foreign GH genes in transgenic mice: (i) regulatory sequences from "housekeeping" genes like the metallothionein (MT) gene (Palmiter *et al.* 1982) or the H2-K^b gene (Morello *et al.* 1986) that are expressed in nearly all tissues, and (ii) regulatory sequences controlling transgene expression restricted to a single tissue or a limited number of tissues (Table 1).

Table 1. Examples of tissue-specific expression of GH fusion genes in transgenic mice

Gene construct	Tissue specificity of expression	Reference
EI-hGH	pancreas	Ornitz *et al.* (1985)
alb-hGH	liver, kidney	Pinkert *et al.* (1987)
PEPCK-bGH	liver, kidney	McGrane *et al.* (1988)
rGH-hGH	pituitary gland	Behringer *et al.* (1988)
osteocalcin-hGH	bone	Baker *et al.* (1992)
rbWAP-hGH	mammary gland, other tissues in some lines	Devinoy *et al.* (1994)
RSV/LTR-bGH	heart, striated muscle	Conti *et al.* (1995)

Regulatory sequences: EI = rat elastase I; alb = mouse albumin; PEPCK = rat phosphoenolpyruvate carboxykinase; rbWAP = rabbit whey acidic protein; RSV/LTR = long terminal repeat of Rous sarcoma virus.

Regulated Expression of GH Transgenes
The expression of different GH fusion genes has been shown to mirror the expression of the genes from which the respective promoter sequences were derived. In MT-GH transgenic mice, transgene expression starts during fetal development and continues throughout the entire life-span (Palmiter *et al.*

1983; Hammer *et al.* 1985). These mice show an increase in fusion gene-specific mRNA and in circulating GH levels following treatment with heavy metals (Palmiter *et al.* 1983). In MT-hGH transgenic mice, increased levels of transgene-specific mRNA in the liver and serum GH concentrations have also been detected following administration of glucocorticoids (Selden *et al.* 1989). A mutated ovine (o) MT promoter has been used to generate transgenic mice in which the expression of the oGH structural gene can be switched on, or entirely turned off, by providing or withholding $ZnSO_4$ in the drinking water (Shanahan *et al.* 1989). Another frequently used promoter for over-expression of heterologous GH genes is derived from the rat phosphoenolpyruvate carboxykinase (PEPCK) gene. Expression of such transgenes starts around birth and can be regulated by diet composition (reduced by high-carbohydrate diet; stimulated by high-protein and/or a low-carbohydrate diet) (McGrane *et al.* 1988). A recent study on PEPCK-bGH transgenic mice demonstrated that vitamin A deficiency caused down-regulation of transgene expression in the liver which could be restored by treatment with all-trans retinoic acid. This suggests that retinoic acid plays a physiological role in regulating hepatic expression of the PEPCK gene (Shin *et al.* 1995).

Unexpected Expression of GH Transgenes

The expression of MT-rGH fusion genes in a specific subset of neurons (which do not normally express endogenous metallothionein or GH) demonstrated for the first time that juxtaposition of sequences from two different genes may result in novel tissue specificities of expression (Swanson *et al.* 1985). A similar phenomenon was observed in whey acidic protein (WAP)-hGH transgenic mice. Although the regulation of transgene expression in the mammary gland mirrored that of the endogenous WAP gene, unexpected expression of the transgene was detected in Bergman glia cells (Günzburg *et al.* 1991).

Phenotypic Alterations in GH Overproducing Mice

Body, Skeletal, Muscle, and Organ Growth

Stimulation of body growth is the most obvious phenotypic effect of GH over-expression in transgenic mice (Palmiter *et al.* 1982; Palmiter *et al.* 1983; Figure 1A). The degree of growth stimulation reached up to twice normal and was similar in transgenic mice carrying hGH, rGH, bGH or oGH sequences. The primary structures of these hormones are different in 1 to 65 out of 191 or 190 amino acids (for review, see Wanke *et al.* 1992).

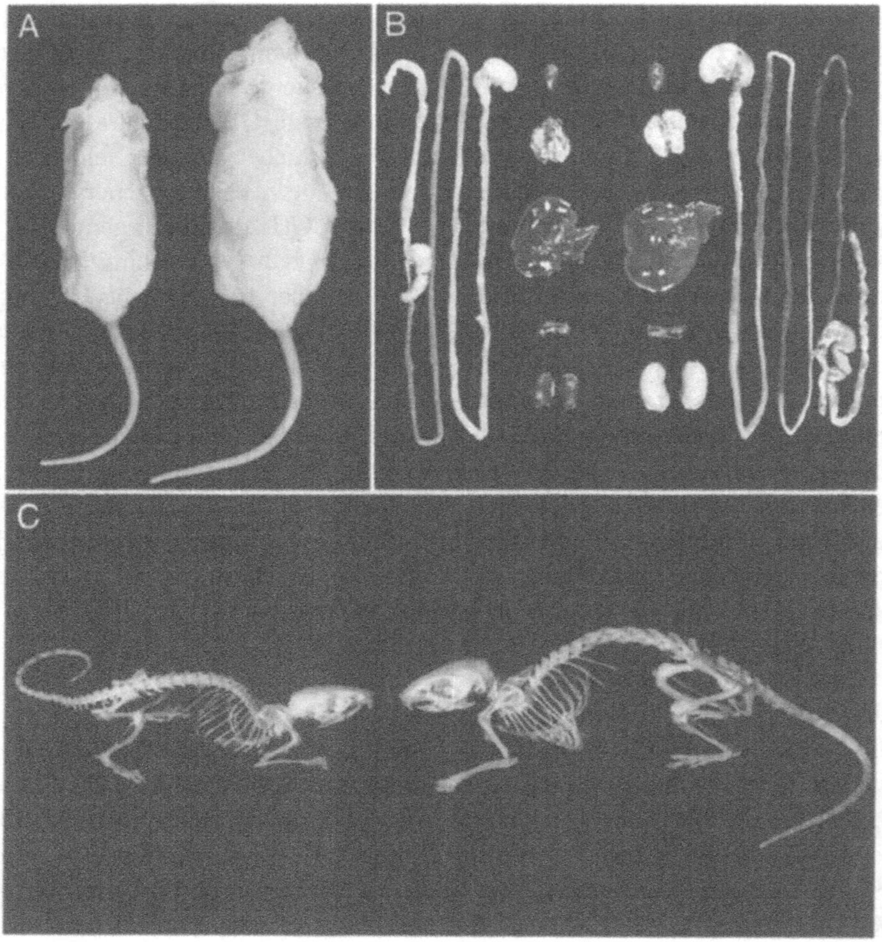

Figure 1: *Growth-promoting effects of hGH overproduction in a MT-hGH transgenic mouse (right) as compared to a non-transgenic littermate control (left). A) Markedly stimulated body growth; B) Visceromegaly which is most prominent in liver and kidneys; C) Disproportionate skeletal gigantism.*

Expression of MT-rGH fusion genes in GH-deficient *lit/lit* mice resulted in an up to four-fold increase in body weight (Hammer *et al.* 1984). Although similar maximum body weights were reached by mice following long-term selection for high body weight, the pattern of their body growth (relatively small increase in body proportions, marked accumulation of fat) was completely different from that of GH transgenic mice (Wolf *et al.* 1991c). A study on oMT-oGH transgenic mice revealed that elevated GH levels increase feed efficiency and lean tissue production (Pomp *et al.* 1992). In addition to altered body composition, changes in the physiological sex-related dimorphism in body growth were noted in some lines of GH transgenic mice. In a line of MT-hGH transgenic mice characterised by very high serum levels of GH, the growth curves and maximum body weights of male and female transgenic mice were almost identical (Brem *et al.* 1989). In this line the lack of sex-related differences in body weight gain may be due to the extremely high levels of serum GH, as well as to the continuous pattern of GH secretion. By contrast, transgenic lines of mice expressing bGH sequences displayed a sex-related difference in body weight. However, at least in older mice, this difference was due to the markedly stimulated growth of the skin of male bGH transgenic mice (Wanke *et al.* 1995b).

In contrast to the generally growth-promoting effects of systemically elevated GH, the expression of the hGH gene under the control of the mouse mammary tumour virus (MMTV) promoter specifically in the cerebral cortex of transgenic mice resulted in a dominant phenocopy of hypopituitarism with growth retardation (Zambrowicz *et al.* 1993). These mice displayed increased hypothalamic somatostatin mRNA levels but decreased GH-releasing hormone (GHRH) mRNA concentrations, demonstrating direct feedback effects by GH on its regulatory peptides.

Analyses of skeletal growth of transgenic mice oversecreting hGH (Wolf *et al.* 1991a), rGH (Shea *et al.* 1990), or oGH (Oberbauer *et al.* 1992) revealed skeletal proportions which were significantly different from those of control mice. An example of disproportionate skeletal gigantism in a MT-hGH transgenic mouse is shown in Figure 1C. Since bone growth is influenced not only by hormones and growth factors but also by physical factors, like mechanical loading, we compared skeletal growth of MT-hGH transgenic mice with that of mice reaching similar maximum body weights as a result of long-term selection for high body weight (Wolf *et al.* 1991b). The fact that bone dimensions of transgenics were as a rule larger than those of weight-selected mice suggested that skeletal gigantism in MT-hGH transgenic mice can be only in part a result of increased body weight. MT-bGH transgenic mice (gonadectomized *vs* sham-operated) have been used to investigate whether sex steroids can regulate bone growth in an animal model with sex steroid-independent GH secretion (Sandstedt *et al.* 1994). Whereas gonadectomy in female mice did not affect body weight gain as well as spinal and tibial growth, both MT-bGH transgenic and control orchidectomized males showed alterations in the proportions between the lumbar spine and the tibia. The authors concluded that the male gonads stimulate pubertal growth of the spine in a GH-independent manner and inhibit tibial growth of adult animals. In transgenic mice with increased levels of circulating GH, skeletal gigantism appears to represent an endocrine effect of GH. However, in a line of osteocalcin-hGH transgenic mice characterised by bone-specific transgene expression bone growth was stimulated with minimal systemic effects suggesting local actions of GH on bone growth (Baker *et al.* 1992). In contrast to GH, increased systemic levels of insulin-like growth factor-I (IGF-I) or IGF-II in transgenic mice did not stimulate bone growth (Mathews *et al.* 1988; Wolf *et al.* 1995b), suggesting that IGFs stimulate bone growth in an auto-paracrine manner rather than in an endocrine manner.

Muscle growth was stimulated by chronic overproduction of GH as evidenced by significantly increased forelimb and hindlimb muscle weights of PEPCK-bGH transgenic mice as compared to controls. However, the increase in muscle weight was less pronounced than the increase in total body weight and was not accompanied by a simultaneous increase in muscle strength (Wolf *et al.* 1995c). Analysis of soleus muscle in MT-hGH transgenic mice revealed elevated muscle weight associated with an increase in cross-sectional areas of both type I and type IIA fibres and a shift in fibre composition towards type I fibres (Dudley and Portanova 1987). Similar findings were reported for PEPCK-bGH transgenic mice (Schenck *et al.* 1992).

Visceromegaly is a common consequence of GH overproduction in transgenic mice (e.g. Figure 1B), however the degree of enlargement varies from organ to organ. Furthermore, differences have been noted between various lines of GH transgenic mice (for review, see Wanke *et al.* 1991b). Allometric analysis of organ growth of both MT-rGH (Shea *et al.* 1987) and MT-bGH transgenic mice (Cecim *et al.* 1993) revealed a special increase in the size of the liver and spleen, whereas kidney growth paralleled total body growth. By contrast, a line of MT-hGH transgenic mice characterised by circulating hGH levels of more than 3000 ng/ml displayed significantly increased weights of heart, lung, spleen, liver and kidneys which were - except for the spleen - also increased relative to body weight (Brem *et al.* 1989). Possible factors responsible for these different results include the species of foreign GH as well as the level of overproduction and have been discussed in detail elsewhere (Brem *et al.* 1989).

Endocrine and Metabolic Changes

Chronic overproduction of GH is associated with increases in the number of hepatic GH and prolactin (PRL) receptors (Orian *et al.* 1991; Aguilar *et al.* 1992), the concentration of GH-binding protein (GHBP) in serum, the amount of bound GH and the proportion of bound GHBP, leading to a massive increase in the levels of GHBP-GH complexes in the circulation (Sotelo *et al.* 1995). Circulating IGF-I concentrations in transgenic mice overexpressing various GH transgenes were reported to be increased by two- to threefold (Palmiter *et al.* 1983; Chow *et al.* 1994). Whereas serum IGFBP-3 levels were increased, circulating IGFBP-2 levels tended to decrease as a consequence of GH over-expression in transgenic mice (Chow *et al.* 1994; Wolf *et al.* 1995a).

As expected from the well-documented feedback control of GH release, persistent ectopic production of GH and elevation of peripheral GH levels in transgenic mice leads to suppression of endogenous GH release with reduced pituitary GHRH binding sites (Sotelo *et al.* 1993). MT-hGH transgenic mice displayed a marked reduction in number and size of somatotrophs but also of lactotrophs, corticotrophs, and gonadotrophs (Stefaneanu *et al.* 1990). By contrast, bGH overproduction in transgenic mice resulted primarily in a reduction of somatotrophs (Stefaneanu *et al.* 1993; Schmidt *et al.* 1994). These differences were explained by the fact that bGH is purely somatotropic, whereas hGH is both somatotropic and lactogenic.

Neuroendocrine and reproductive consequences of expression of various GH transgenes in mice have recently been summarised (Bartke *et al.* 1994) and are therefore not discussed in detail in the present review. Although a number of studies on female fertility of GH transgenic mice have been published, no clear view has emerged of the effects exerted by GHs from different species. An early study on female MT-hGH transgenic mice characterised by relatively low levels of circulating hGH (3.8±0.3 ng/ml) revealed infertility due to luteal failure which was caused by insufficient secretion of prolactin from the anterior pituitary gland (Bartke *et al.* 1988). The latter has been interpreted as a consequence of the lactogenic action of hGH in mice, resulting in functional hyperprolactinaemia and suppression of endogenous PRL release. However, the same group reported normal fertility of females from a line of PEPCK-hGH transgenic mice characterised by circulating GH concentrations between 100 and 700 ng/ml (Milton *et al.* 1992). Our own MT-hGH transgenic female mice exhibiting serum GH levels of more than 3000 ng/ml were almost exclusively infertile. However, transgenic offspring were obtained following transcaryotic implantation of transgenic ovaries into normal mice (Brem *et al.* 1990).

In spite of marked systemic effects of elevated GH, female transgenic mice expressing bGH or oGH fusion genes were initially reported to be fertile (Mayerhofer *et al.* 1990; Orian *et al.* 1989). By contrast, more recent studies revealed severe impairment of female reproductive functions as a result of overexposure to bGH or oGH and it was suspected that this infertility was due to luteal failure (Cecim *et al.* 1995; Pomp *et al.* 1995). These different results indicate that the effects of GH excess on female reproductive functions may be dose-related and/or specific for the respective line investigated.

In contrast to females, most transgenic male mice expressing heterologous GH genes are fertile (for review, see Bartke *et al.* 1994). However, quantitative analysis of reproductive performance revealed inferior breeding performance of transgenic vs. control males in terms of interval between pairing and reproductive life-span. Reproductive deficits in GH transgenic males did not appear to be due to suppression of spermatogenesis or of androgen production, but to changes in copulatory behaviour (Bartke *et al.* 1992).

The effects of GH overproduction on carbohydrate metabolism have been subject to several studies. Transgenic mice expressing hGH or bGH coding sequences under the control of the MT promoter exhibited significantly increased fasting insulin levels, whereas glucose levels were not significantly different from those of controls (for review, see Wanke *et al.* 1992). Hyperinsulinaemia in MT-bGH transgenic mice was associated with an overproportionate increase in total B cell volume of islets of Langerhans (Wanke *et al.* 1994). In contrast to MT-bGH transgenic mice, expression of the bGH coding sequence under the control of the PEPCK promoter was reported to cause hyperinsulinaemia only in lines with very high levels (mean = 2.5 μg/ml) of circulating bGH (McGrane *et al.* 1990). Differences in glucose homeostasis between the various lines of GH transgenic mice have been interpreted as results of differences in tissue specificity, onset of expression, and hormonal regulation of GH transgene expression (for review, see Wanke *et al.* 1992). The lipid status of MT-hGH transgenic mice was characterised by a significant increase in serum cholesterol levels, whereas no significant changes were observed for serum levels of triglycerides (Brem *et al.* 1989).

Ectopic expression of hGH or bGH genes in transgenic mice was associated with a significant increase in plasma corticosterone levels. The degree of elevation of plasma corticosterone did not correlate with male or female fertility or with life expectancy (Cecim *et al.* 1991).

Pathomorphological Changes

Renal Systemic overproduction of GH in transgenic mice - either due to ectopic expression of heterologous GH genes or to stimulation of endogenous GH release by expression of a GH-releasing hormone transgene - regularly leads to a typical spectrum of renal alterations. These lesions were first reported in 1988 in MT-hGH (Brem and Wanke 1988) and MT-bGH transgenic mice (Doi *et al.* 1988) and subsequently found in many different lines of GH transgenic mice (for review, see Wanke *et al.* 1991b; Wanke *et al.* 1992). Initial changes include glomerular hypertrophy and progressive glomerulosclerosis, whereas end-stage renal lesions are characterised by a marked atrophy of nephrons, sclerosis and hyalinosis of the remaining glomeruli and pronounced tubulocystic alterations (Figure 2). Both, morphological and clinico-chemical investigations (Brem *et al.* 1989; Quaife *et al.* 1989) indicated renal failure to be the primary cause of shortened life-span of GH transgenic mice (Wolf *et al.* 1993).

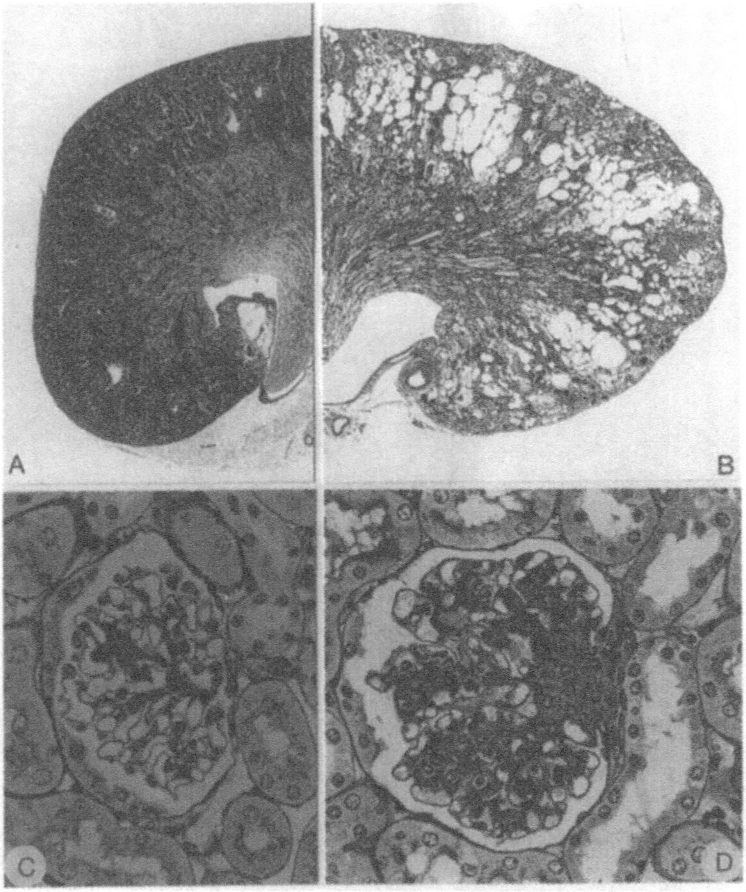

Figure 2: *Renal lesions resulting from GH overproduction in GH transgenic mice. A) Kidney of a control mouse; B) Cystic tubular dilation, numerous proteinaceous casts in a kidney of a 12-week-old MT-hGH transgenic mouse; C) Glomerulus of a control mouse; D) Glomerular hypertrophy and sclerosis in a 24-week-old MT-hGH transgenic mouse.*

Hepatic In addition to renal lesions, hepatic alterations have been consistently observed in GH transgenic mice. In GH over-producing mice, the liver shows a marked enlargement which was found to be overproportionate to general size increase. In addition, a typical sequence of histological findings was reported. Initial alterations were characterised by centrolobular

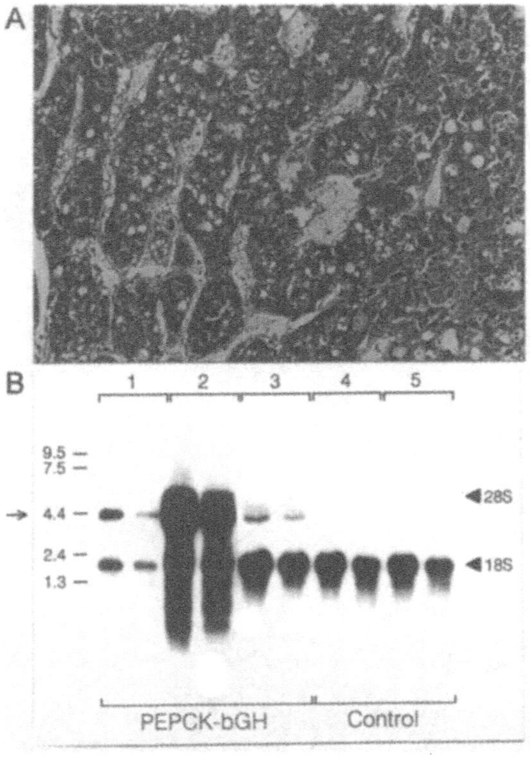

Figure 3: *Hepatocarcinogenesis in GH transgenic mice. A) Trabecular hepatocellular carcinoma, 13-month-old PEPCK-bGH transgenic mouse; B) Re-expression of the IGF-II gene during hepatocarcinogenesis in a 13-month-old PEPCK-bGH transgenic mouse. RNA was extracted from liver tissue and subjected to Northern blot analysis using an IGF-II-specific ^{32}P-labeled complementary RNA probe (Wolf et al. 1994). Samples 1 and 2 were derived from a hepatocellular carcinoma; sample 3 was obtained from macroscopically normal liver tissue of the GH transgenic mouse; samples 4 and 5 were derived from normal liver tissue of a control mouse. The characteristic 4.4 kb IGF-II transcript is marked by an arrow. The relative mobility of the 18S and 28S ribosomal RNA is indicated at the right side of the blot.*

hepatocellular and hepatonuclear hypertrophy. The next stage saw massive diffuse enlargement associated with extreme pleomorphism of the liver cells and their nuclei as well as frequently occurring nuclear pseudoinclusions. In addition to hypertrophic changes, regressive and regenerative changes were observed. Apart from non-neoplastic liver changes, hepatocellular tumours including both adenoma and carcinoma (Figure 3A) occurred in an age-dependent fashion and with a very high frequency in transgenic mice expressing oGH or bGH fusion genes (Orian *et al.* 1990; Wanke *et al.*

1991a). Tumourigenesis of the liver of bGH transgenic mice was associated with re-expression of the IGF-II gene (Wanke *et al.* 1995a; Figure 3B).

Other Other alterations which have been observed as a consequence of GH overproduction in transgenic mice include multifocal myocardial fibrosis (associated with advanced renal changes), extramedullary hematopoiesis in the spleen, and vascular remodelling (for review, see Wanke *et al.* 1991b). Mammary adenocarcinomas have been observed in MT-hGH transgenic mice with a hybrid genetic background (C57Bl/6 x DBA/2 or C57Bl/6 x C3H) (Tornell *et al.* 1991; Steger *et al.* 1993).

Animal Models Based on GH Overproduction in Transgenic Mice

Long-term Effects of Elevated GH Levels

The availability of recombinant GH and the development of long-acting formulations led to a wide-spread medical use of GH including paediatric (Hindmarsh *et al.* 1991) and geriatric indications (Rudman *et al.* 1990). Furthermore, GH is used by athletes like bodybuilders to increase muscle mass and strength (McIntyre 1987; Wagner 1991). On the other hand, there is recent concern about ethical and medical risks especially in treating patients that do not meet the classical symptoms of GH deficiency (Lantos *et al.* 1989; Schoen 1991). Therefore, there is urgent need for suitable models to evaluate long-term benefits and risks of elevated GH levels.

Treatment of rodents with heterologous GH does not seem to be an appropriate model, since (i) the animals may be disturbed by repeated administrations, (ii) the biological effect of GH may depend on the mode of administration (Maiter *et al.* 1988), and (iii) only short-term studies are possible due to a marked immunological response to the administered foreign GH (for review, see Wolf *et al.* 1993). The latter was also observed in a transcaryotic gene therapy experiment, in which rats were grafted with autologous fibroblasts genetically engineered to synthesise hGH (Chang *et al.* 1990).

A different approach of creating a situation of exposure to increased levels of GH is the transplantation of functionally active pituitary tumours, which has been performed in rats (Gilbert *et al.* 1985; Mori *et al.* 1986). However, as a rule these tumours do not secrete GH exclusively, but secrete other pituitary peptides in addition. Furthermore, the alterations observed in grafted rats cannot be attributed solely to the action of GH and may reflect the effects of the tumour.

By contrast, over-expression of GH genes in transgenic mice provides a unique opportunity to study the effects of long-term elevated circulating GH. As these mice may synthesise heterologous GH as an authentic self-molecule, chronic overexposure to this GH is not disturbed by immunological complications. In addition, the timing, level, and tissue specificity of GH overproduction can be adapted to the respective experimental question by using suitable regulatory sequences to drive transgene expression and by selecting appropriate lines of transgenic mice. Whereas lines expressing low levels of heterologous GH may be considered as models for a therapeutic use of GH, transgenic mice with a high expression of GH fusion genes may help to show the adverse effects of GH more clearly and at an earlier age.

Glomerulosclerosis

The term "glomerulosclerosis" is commonly used to designate glomerular scarring, irrespective of the nature of the process and the composition of the scar, and refers to a variety of glomerular lesions including increases in mesangial matrix, collapse and obliteration of capillary lumina, accumulation of hyaline, and synechial attachments to Bowman's capsule (Schwartz and Lewis 1985; Floege *et al.* 1992; Striker 1993). Glomerulosclerosis may be focal, involving some glomeruli, or it may be diffuse, involving all or nearly all (>80%) glomeruli, and may be segmental, involving a portion of the glomerulus, or may be global, involving the whole glomerulus (Churg and Sobin 1982).

In human disease, glomerulosclerosis can occur in a primary form (primary focal and segmental glomerulosclerosis), the condition associated with the idiopathic nephrotic syndrome, or it may arise in association with a variety of systemic or renal diseases (secondary glomerulosclerosis) (Goldszer *et al.* 1984). Systemic diseases which may involve the development of glomerulosclerosis can be grouped into broad categories, such as vascular diseases (e.g. essential hypertension), metabolic disorders (e.g. diabetes mellitus), hereditary diseases (e.g. Fabry's disease), and systemic immunological diseases (e.g. systemic lupus erythematosus). Primary immune glomerular injury syndromes, such as membranous nephropathy, mesangioproliferative glomerulonephritis, membranoproliferative glomerulonephritis or crescentic glomerulonephritis can all evolve into diffuse glomerulosclerosis (Klahr *et al.* 1988; Schreiner 1990). Glomerulosclerosis may also develop in patients with primary tubulointerstitial renal diseases, such as reflux nephropathy or analgesic nephropathy, to name but a few (Goldszer *et al.* 1984). Chronic renal insufficiency, once established, tends to progress to end-stage kidney failure which is characterised by progressive glomerulosclerosis independent of the nature of the initiating nephropathy (El Nahas

1989). Thus, there is urgent need for suitable animal models to clarify factors involved in the pathogenesis of glomerulosclerosis.

Glomerulosclerosis in GH transgenic mice develops regularly without any exogenous treatment and is characterised by all the typical lesions listed above. Since treatment strategies can already be tested in a stage of the disease in which the animals do not suffer, the welfare of GH transgenic mice is not necessarily compromised in such studies.

By contrast, subtotal nephrectomy, which is the most widely used approach to induce glomerulosclerosis and renal failure, causes a very extreme situation for the animals (Elema *et al.* 1988). Four post-surgical stages can be identified: (i) acute renal failure (which causes early mortality, the rate of which depends on amount of tissue removed, the animal's age and diet), (ii) improvement of renal function, (iii) stable renal insufficiency, and (iv) deterioration of kidney function leading to end-stage renal disease (Kleinknecht *et al.* 1988). Subtotal nephrectomy may be obtained by uninephrectomy and resection of poles of the remnant kidney or by uninephrectomy and ligation of branches of the contralateral renal artery. Although both methods are used they are not at all equivalent, since ligation of branches of the renal artery causes immediate severe hypertension whereas hypertension occurs only late after pole resection. Another reason for the limited standardisation of this model is the variable compensatory degree of hypertrophy of the remnant kidney (Kleinknecht *et al.* 1995).

In addition to surgical methods, chemicals have been used to induce glomerulosclerosis in laboratory animals. Treatment with puromycin aminonucleoside results in a nephrotic syndrome with variable incidence of glomerulosclerosis. A similar nephrotic syndrome is produced by intravenous injection of Adriamycin, an antineoplastic antibiotic. Adriamycin toxicity causes high long-term mortality with mortality rates of up to 40% even after low-dose treatment (Bertani *et al.* 1986). Both models are not standardised and therefore exhibit the same problems as the nephrectomy models (Strauch and Gretz 1988).

Various approaches have been used to establish rat models for diabetic nephropathy, with the streptozotocin (STZ)-treated rat being the most widely used (Strauch and Gretz 1988). "Glomerulosclerosis" has been observed in these rats after 9 to 27 months of diabetes induced by STZ treatment. However, these alterations were not sufficient to cause any significant capillary occlusion (Hasslacher 1988). Therefore, the STZ diabetic rat is not useful as a model for advanced diabetic glomerulopathy (Hirose *et al.* 1982). By contrast, glomerular alterations in GH transgenic mice mimic those in human diabetes mellitus (Doi *et al.* 1988). In addition, tubulocystic alterations occurring in GH transgenic mice may represent a model for acquired cystic kidney disease (Grantham 1991).

Hepatocarcinogenesis

Hepatocarcinogenesis is a complex process involving a defined sequence of pathological alterations (for review, see Farber and Sarma 1987). Transgenic mice over-producing ovine (Orian *et al.* 1990) or bovine GH (Wanke *et al.* 1991a) have been suggested as a promising model for understanding the multistep nature of liver carcinogenesis. The advantages of GH transgenic mouse models over other tumour models may include: (i) the absence of protocols that involve large doses of potent carcinogens, (ii) the ease of producing additional mice once breeding lines have been established, and (iii) the possibility of examining effects of cofactors under well controlled conditions (Orian *et al.* 1990). Based on the assumption that GH transgenic mice are primed for hepatocellular carcinoma by an apparently non-mutagenic effect, Orian *et al.* (1990) suggested these mice as an important model for carcinogen testing with a presumably greatly shortened latency of liver tumour formation. From the animal welfare aspect, it has to be taken into account that spontaneous neoplastic alterations of the liver of GH transgenic mice occur relatively late in life (adenomas from 6 months of age and carcinomas from 12 months of age, respectively). At this age, GH transgenic mice may already be compromised by consequences of the simultaneously occurring kidney lesions. Nevertheless, given that GH is a promoter of tumourigenesis, young GH transgenic mice could be helpful for carcinogen testing without suffering especially if preneoplastic alterations are evaluated. Since - like many other models for hepatotumourigenesis (Cariani *et al.* 1991; Schirmacher *et al.* 1992) - reactivation of IGF-II gene expression has been observed in neoplastic liver tissue from GH transgenic mice, these mice may help to clarify the role of IGF-II in hepatocarcinogenesis.

Welfare Considerations

Over-expression of GH genes in transgenic mice causes an age-dependent spectrum of endocrine and metabolic changes as well as severe pathological changes of the liver and kidneys, the latter being the primary cause of shortened life-span of these animals. Therefore, alterations observed in GH transgenic mice have been used as an example to illustrate general welfare problems with regard to transgenic animals (Poole 1995). However, GH transgenic mice provide unique models for evaluating long-term effects of over-exposure to GH, which cannot be done in classical rodent models. GH transgenic mice have been accepted as a model for studying the pathogenesis of glomerulosclerosis. Glomerulosclerosis occurs reproducibly without any surgical intervention or other exogenous treatments. Most importantly, therapeutic strategies can be tested before the animals suffer from the consequences of renal insufficiency. In this aspect, the welfare of GH transgenic mice is clearly less disturbed than that of the models produced by subtotal nephrectomy. Furthermore, GH transgenic mice have been suggested as a non-genotoxic model for hepato-carcinogenesis, although this model situation seems to be less well defined than the ones mentioned before. All these scientific questions currently investigated by using GH transgenic mice appear important for mankind and can be justified on ethical criteria.

Acknowledgements

We would like to thank Prof. Dr. G. Brem and Prof. Dr. W. Hermanns for continuous support; and Dr. I. Renner-Müller, Dr. A. Höflich, K. Brodowski, and A. Blackburn for helpful comments on the manuscript. Our studies were supported by the Deutsche Forschungsgemeinschaft (He 1731/1-4) and by the Studienstiftung des Deutschen Volkes.

References

Aguilar RC, Fernandez HN, Dellacha JM, Calandra RS, Bartke A, Ghosh PK, Turyn D (1992) Somatotropic and lactotropic receptors in transgenic mice expressing human or bovine growth hormone genes. Transgenic Research 1, 221-227.

Baker AR, Hollingshead PG, Pitts-Meek S, Hansen S, Taylor R, Stewart TA (1992) Osteoblast-specific expression of growth hormone stimulates bone growth in transgenic mice. Molecular and Cellular Biology 12, 5541-5547.

Bartke A, Steger RW, Hodges SL, Parkening TA, Collins TJ, Yun JS, Wagner TE (1988) Infertility in transgenic female mice with human growth hormone expression: evidence for luteal failure. Journal of Experimental Zoology 248, 121-124.

Bartke A, Naar EM, Johnson L, May MR, Cecim M, Yun JS, Wagner TE (1992) Effects of expression of human or bovine growth hormone genes on sperm production and male reproductive performance in four lines of transgenic mice. Journal of Reproduction and Fertility 95, 109-118.

Bartke A, Cecim M, Tang K, Steger RW, Chandrashekar V, Turyn D (1994) Neuroendocrine and reproductive consequences of overexpression of growth hormone in transgenic mice. Proceedings of the Society for Experimental Biology and Medicine 206, 345-359.

Behringer RR, Mathews LS, Palmiter RD, Brinster RL (1988) Dwarf mice produced by genetic ablation of growth hormone-expressing cells. Genes & Development 2, 453-461.

Bertani T, Rocchi G, Sacchi G, Mecca G, Remuzzi G (1986) Adriamycin-induced glomerulo-sclerosis in the rat. American Journal of Kidney Disease 7, 12-19.

Brem G, Wanke R, Wolf E, Buchmüller T, Müller M, Brenig B, Hermanns W (1989) Multiple consequences of human growth hormone expression in transgenic mice. Molecular Biology and Medicine 6, 531-547.

Brem G, Baunack E, Müller M, Winnacker EL (1990) Transgenic offspring by transcaryotic implantation of transgenic ovaries into normal mice. Molecular Reproduction and Development 25, 42-44.

Brem G (1993) Transgenic animals. In: Biotechnology (Rehm HJ, Reed G, eds). Weinheim: VCH, pp 745-832.

Brem G, Wanke R (1988) Phenotypic and pathomorphological characteristics in a half-sib-family of transgenic mice carrying foreign MT-hGH genes. In: New Developments in Biosciences: Their Implications for Laboratory Animal Science (Beynen AC, Solleveld HA, eds). Dordrecht: Martinus Nijhoff Publishers, pp 93-98.

Bry L, Falk P, Huttner K, Ouellette A, Midtvedt T, Gordon JI (1994) Paneth cell differentiation in the developing intestine of normal and transgenic mice. Proceedings of the National Academy of Sciences of the USA 91, 10335-10339.

Cariani E, Dubois N, Lasserre C, Briand P, Brechot C (1991) Insulin-like growth factor II (IGF-II) mRNA expression during hepatocarcinogenesis in transgenic mice. Journal of Hepatology 13, 220-226.

Cecim M, Ghosh PK, Esquifino AI, Began T, Wagner TE, Yun JS, Bartke A (1991) Elevated corticosterone levels in transgenic mice expressing human or bovine growth hormone genes. Neuroendocrinology 53, 313-316.

Cecim M, Bartke A, Yun JG, Wagner TE (1993) Growth allometry of transgenic mice expressing the mouse metallothionein-I/bovine growth hormone gene. Transgene 1, 125-132.

Cecim M, Kerr J, Bartke A (1995) Infertility in transgenic mice overexpressing the bovine growth hormone gene: luteal failure secondary to prolactin deficiency. Biology of Reproduction 52, 1162-1166.

Chang PL, Capone JP, Brown GM (1990) Autologous fibroblast implantation - feasibility and potential problems in gene replacement therapy. Molecular Biology and Medicine 7, 461-470.

Chen WY, Chen NY, Yun J, Wagner TE, Kopchick JJ (1994) In vitro and in vivo studies of antagonistic effects of human growth hormone analogs. Journal of Biological Chemistry 269, 15892-15897.

Chen WY, Chen NY, Yun J, Wight DC, Wang XZ, Wagner TE, Kopchick JJ (1995) Amino acid residues in the third alpha-helix of growth hormone involved in growth promoting activity. Molecular Endocrinology 9, 292-302.

Chow JC, Murray JD, Pomp D, Baldwin RL, Calvert CC, Oberbauer AM (1994) Regulation of insulin-like growth factor-I and binding protein-3 expression in oMtla-oGH transgenic mice. Transgenic Research 3, 127-133.

Churg J, Sobin LH (1982) Renal Disease: Classification and Atlas of Glomerular Diseases. Tokyo, New York: Igaku-Shoin.

Conti FG, Powell R, Pozzi L, Zezze G, Faraggiana T, Gannon F, Fabbrini A (1995) A novel line of transgenic mice (RSV/LTR-bGH) expressing growth hormone in cardiac and striated muscle. Growth Regulation 5, 101-108.

Devinoy E, Thepot D, Stinnakre MG, Fontaine ML, Grabowski H, Puissant C, Pavirani A, Houdebine LM (1994) High level production of human growth hormone in the milk of trans-genic mice: the upstream region of the rabbit whey acidic protein (WAP) gene targets transgene expression to the mammary gland. Transgenic Research 3, 79-89.

Doi T, Striker LJ, Quaife C, Conti FG, Palmiter R, Behringer R, Brinster R, Striker GE (1988) Progressive glomerulosclerosis develops in transgenic mice chronically expressing growth hormone and growth hormone releasing factor but not in those expressing insulinlike growth factor-1. American Journal of Pathology 131, 398-403.

Dudley GA, Portanova R (1987) Histochemical characteristics of soleus muscle in hGH transgenic mice. Proceedings of the Society for Experimental Biology and Medicine 185, 403-408.

Ehrlein J, Wanke R, Weis S, Brem G, Hermanns W (1994) Sensitive detection of human growth hormone mRNA in routinely formalin-fixed, paraffin-embedded transgenic mouse tissues by non-isotopic in situ hybridization. Histochemistry 102, 145-152.

El Nahas AM (1989) Glomerulosclerosis: insights into pathogenesis and treatment. Nephrology, Dialysis and Transplantation 4, 843-853.

Elema JD, Weening JJ, Grond J (1988) Focal glomerular hyalinosis and sclerosis in amino-nu-cleoside and adriamycin nephrosis: pathogenetic and therapeutic considerations. Contributions to Nephrology 60, 73-82.

Farber E, Sarma DSR (1987) Hepatocarcinogenesis: a dynamic cellular perspective. Laboratory Investigation 56, 4-22.

Floege J, Burns NW, Alpers CE, Yoshimura A, Pritzl P, Gordon K, Seifert RA, Bowen-Pope DF, Couser WG, Johnson RJ (1992) Glomerular cell proliferation and PDGF expression precede glomerulosclerosis in the remnant kidney model. Kidney International 41, 297-309.

Gilbert PL, Siegel RJ, Melmed S, Sherman CT, Fishbein MC (1985) Cardiac morphology in rats with growth hormone-producing tumours. Journal of Molecular and Cellular Cardiology 17, 805-811.

Goldszer RC, Sweet J, Cotran RS (1984) Focal segmental glomerulosclerosis. Annual Review of Medicine 35, 429-449.

Grantham JJ (1991) Acquired cystic kidney disease. Kidney International 40, 143-152.

Günzburg WH, Salmons B, Zimmermann B, Müller M, Erfle V, Brem G (1991) A mammary-specific promoter directs expression of growth hormone not only to the mammary gland, but also to Bergman glia cells in transgenic mice. Molecular Endocrinology 5, 123-133.

Hammer RE, Palmiter RD, Brinster RL (1984) Partial correction of murine hereditary growth disorder by germ-line incorporation of a new gene. Nature 311, 65-67.

Hammer RE, Brinster RL, Palmiter RD (1985) Use of gene transfer to increase animal growth. Cold Spring Harbor Symposia on Quantitative Biology 50, 379-387.

Hasslacher C (1988) Development of diabetic nephropathy in the rat: morphological, functional, and pathophysiological aspects. Contributions to Nephrology 60, 9-17.

Hindmarsh PC, Bridges NA, Brook CGD (1991) Wider indications for treatment with biosynthetic human growth hormone in children. Clinical Endocrinology (Oxford) 34, 417-427.

Hirose K, Osterby R, Nozawa M, Gundersen HJG (1982) Development of glomerular lesions in experimental long-term diabetes in the rat. Kidney International 21, 689-695.

Klahr S, Schreiner G, Ichikawa I (1988) The progression of renal disease. The New England Journal of Medicine 318, 1657-1666.

Kleinknecht C, Laouari D, Burtin M (1988) Uremic rat model: experience with young rats. Contributions to Nephrology 60, 27-38.

Kleinknecht C, Terzi F, Burtin M, Laouari D, Maniar S (1995) Experimental models of nephron reduction: some answers, many questions. Kidney International Supplement 49, 51-54.

Lantos J, Siegler M, Cuttler L (1989) Ethical issues in growth hormone therapy. Journal of the American Medical Association 261, 1020-1024.

Maiter D, Underwood LE, Maes M, Davenport ML, Ketelslegers JM (1988) Different effects of intermittent and continuous growth hormone (GH) administration on serum somatomedin-C/insulin-like growth factor I and liver GH receptors in hypophysectomized rats. Endocrinology 123, 1053-1059.

Mathews LS, Hammer RE, Behringer RR, D'Ercole AJ, Bell GI, Brinster RL, Palmiter RD (1988) Growth enhancement of transgenic mice expressing human insulin like growth factor I. Endocrinology 123, 2827-2833.

Mayerhofer A, Weis J, Bartke A, Wagner TE (1990) Effects of transgenes for human and bovine growth hormones on age-related changes in ovarian morphology in mice. Anatomical Record 227, 175-186.

McGrane MM, de Vente J, Yun J, Bloom J, Park E, Wynshaw-Boris A, Wagner T, Rottman FM, Hanson RW (1988) Tissue-specific expression and dietary regulation of a chimeric phosphoenolpyruvate carboxykinase/bovine growth hormone gene in transgenic mice. Journal of Biological Chemistry 263, 11443-11451.

McGrane MM, Yun JS, Moorman AFM, Lamers WH, Hendrick GK, Arafah BM, Park EA, Wagner TE, Hanson RW (1990) Metabolic effects of developmental, tissue-, and cell-specific expression of a chimeric phosphoenolpyruvate carboxykinase (GTP)/bovine growth hormone gene in transgenic mice. Journal of Biological Chemistry 265, 22371-22379.

McIntyre JG (1987) Growth hormone and athletes. Sports Medicine 4, 129-142.

Milton S, Cecim M, Li YS, Yun JS, Wagner TE, Bartke A (1992) Transgenic female mice with high human growth hormone levels are fertile and capable of normal lactation without having been pregnant. Endocrinology 131, 536-538.

Morello D, Moore G, Salmon AM, Yaniv M, Babinet C (1986) Studies on the expression of an H-2K/human growth hormone fusion gene in giant transgenic mice. EMBO Journal 5, 1877-1883.

Mori H, Yamashita H, Nakanishi C, Koizumi K, Makino S, Kishimoto Y, Hayashi Y (1986) Proteinuria induced by transplantable rat pituitary tumor MtT SA5. Laboratory Investigation 54, 636-644.

Oberbauer AM, Currier TA, Nancarrow CD, Ward KA, Murray JD (1992) Linear bone growth of oMtla-oGH transgenic male mice. American Journal of Physiology 262, E936-E942.

Orian JM, Lee CS, Weiss LM, Brandon MR (1989) The expression of a metallothionein-ovine growth hormone fusion gene in transgenic mice does not impair fertility but results in pathological lesions in the liver. Endocrinology 124, 455-463.

Orian JM, Tamakoshi K, Mackey IR, Brandon MR (1990) New murine model for hepatocellular carcinoma: transgenic mice expressing metallothionein-ovine growth hormone fusion genes. Journal of the National Cancer Institute 82, 393-398.

Orian JM, Snibson K, Stevenson JL, Brandon MR, Herington AC (1991) Elevation of growth hormone (GH) and prolactin receptors in transgenic mice expressing ovine GH. Endocrinology 128, 1238-1246.

Ornitz DM, Palmiter RD, Hammer RE, Brinster RL, Swift GH, MacDonald RJ (1985) Specific expression of an elastase-human growth hormone fusion gene in pancreatic acinar cells of transgenic mice. Nature 313, 600-602.

Palmiter RD, Brinster RL, Hammer RE, Trumbauer ME, Rosenfeld MG, Birnberg NC, Evans RM (1982) Dramatic growth of mice that develop from eggs microinjected with metallothionein-growth hormone fusion genes. Nature 300, 611-615.

Palmiter RD, Norstedt G, Gelinas RE, Hammer RE, Brinster RL (1983) Metallothionein-human GH fusion genes stimulate growth of mice. Science 222, 809-814.

Pinkert CA, Ornitz DM, Brinster RL, Palmiter RD (1987) An albumin enhancer located 10 kb upstream functions along with its promoter to direct efficient, liver-specific expression in transgenic mice. Genes & Development 1, 268-276.

Pomp D, Nancarrow CD, Ward KA, Murray JD (1992) Growth, feed efficiency and body composition of transgenic mice expressing a sheep-metallothionein 1a-sheep growth hormone fusion gene. Livestock Production Science 31, 335-350.

Pomp D, Geisert RD, Durham CM, Murray JD (1995) Rescue of pregnancy and maintenance of corpora lutea in infertile transgenic mice expressing an ovine metallothionein 1a-ovine growth hormone fusion gene. Biology of Reproduction 52, 170-178.

Poole TB (1995) Welfare considerations with regard to transgenic animals. Animal Welfare 4, 81-85.

Quaife CJ, Mathews LS, Pinkert CA, Hammer RE, Brinster RL, Palmiter RD (1989) Histopathology associated with elevated levels of growth hormone and insulin-like growth factor I in transgenic mice. Endocrinology 124, 40-48.

Rudman D, Feller AG, Nagraj HS, Gergans GA, Lalitha PY, Goldberg AF, Schlenker RA, Cohn L, Rudman IW, Mattson DE (1990) Effects of human growth hormone in men over 60 years old. The New England Journal of Medicine 323, 1-6.

Sandstedt J, Ohlsson C, Norjavaara E, Nilsson J, Tornell J (1994) Disproportional bone growth and reduced weight gain in gonadectomized male bovine growth hormone transgenic and normal mice. Endocrinology 135, 2574-2580.

Schenck E, Weis S, Wanke R, Brem G, Hermanns W (1992) Morphometric evaluation of muscle in growth hormone transgenic mice. Clinical Neuropathology 11, 172-173 (Abstract).

Schirmacher P, Held WA, Yang D, Chisari FV, Rustum Y, Rogler CE (1992) Reactivation of insulin-like growth factor II during hepatocarcinogenesis in transgenic mice suggests a role in malignant growth. Cancer Research 52, 2549-2556.

Schmidt P, Haider I, Weis S, Brem G, Dahme E (1994) Morphometry of the pituitary gland of growth hormone-transgenic mice. Acta Stereologica 13, 15-20.

Schoen EJ (1991) Growth hormone in youth and old age. Journal of the American Geriatric Society 39, 839.

Schreiner GF (1990) Pathways leading from glomerular injury to glomerulosclerosis. Contributions to Nephrology 86, 1-18.

Schwartz MM, Lewis EJ (1985) Focal segmental glomerular sclerosis: the cellular lesion. Kidney International 28, 968-974.

Selden RF, Yun JS, Moore DD, Rowe ME, Malia MA, Wagner TE, Goodman HM (1989) Gluco-corticoid regulation of human growth hormone expression in transgenic mice and transiently trans-fected cells. Journal of Endocrinology 122, 49-60.

Shanahan CM, Rigby NW, Murray JD, Marshall JT, Townrow CA, Nancarrow CD, Ward KA (1989) Regulation of expression of a sheep metallothionein 1a-sheep growth hormone fusion gene in transgenic mice. Molecular and Cellular Biology 9, 5473-5479.

Shea BT, Hammer RE, Brinster RL (1987) Growth allometry of the organs in giant transgenic mice. Endocrinology 121, 1924-1930.

Shea BT, Hammer RE, Brinster RL, Ravosa MR (1990) Relative growth of the skull and post-cranium in giant transgenic mice. Genetical Research, Cambridge 56, 21-34.

Shin DJ, Tao A, McGrane MM (1995) Effects of vitamin A deficiency and retinoic acid treatment on expression of a phosphoenolpyruvate carboxykinase-bovine growth hormone gene in transgenic mice. Biochemical and Biophysical Research Communications 213, 706-714.

Sotelo AI, Bartke A, Turyn D (1993) Effects of bovine growth hormone (GH) expression in transgenic mice on serum and pituitary immunoreactive mouse GH levels and pituitary GH-releasing factor binding sites. Acta Endocrinologica (Copenhagen) 129, 446-452.

Sotelo AI, Dominici FP, Engbers C, Bartke A, Talamantes F, Turyn D (1995) Growth hormone-binding protein in normal mice and in transgenic mice expressing bovine growth hormone gene. American Journal of Physiology 268, E745-E751.

Stefaneanu L, Kovacs K, Horvath E, Losinski NE, Mayerhofer A, Wagner TE, Bartke A (1990) An immunocytochemical and ultrastructural study of adenohypophyses of mice transgenic for human growth hormone. Endocrinology 126, 608-615.

Stefaneanu L, Kovacs K, Bartke A, Mayerhofer A, Wagner TE (1993) Pituitary morphology of transgenic mice expressing bovine growth hormone. Laboratory Investigation 68, 584-591.

Steger RW, Bartke A, Cecim M (1993) Premature ageing in transgenic mice expressing different growth hormone genes. Journal of Reproduction and Fertility, supplement 46, 61-75.

Strauch M, Gretz N (1988) Animal models to induce renal failure: a historical survey. Contributions to Nephrology 60, 1-8.

Striker LJ (1993) Modern renal biopsy interpretation: can we predict glomerulosclerosis. Seminars in Nephrology 13, 508-515.

Swanson LW, Simmons DM, Arriza J, Hammer R, Brinster R, Rosenfeld MG, Evans RM (1985) Novel developmental specificity in the nervous system of transgenic animals expressing growth hormone fusion genes. Nature 317, 363-366.

Tornell J, Rymo L, Isaksson OG (1991) Induction of mammary adenocarcinomas in metallothionein promoter-human growth hormone transgenic mice. International Journal of Cancer 49, 114-117.

Wagner EF, Covarrubias L, Stewart TA, Mintz B 1983 Prenatal lethalities in mice homozygous for human growth hormone gene sequences integrated in the germ line. Cell 35, 647-655.

Wagner JC (1991) Enhancement of athletic performance with drugs. An overview. Sports Medicine 12, 250-265.

Wanke R, Folger S, Hermanns W, Wolf E, Schams D, Brem G (1991a) Induktion neoplastischer und nicht-neoplastischer Leberveränderungen durch Wachstumshormon-Überproduktion bei bGH-transgenen Mäusen. Verhandlungen der Deutschen Gesellschaft für Pathologie 75, 312.

Wanke R, Hermanns W, Folger S, Wolf E, Brem G (1991b) Accelerated growth and visceral lesions in transgenic mice expressing foreign genes of the growth hormone family. An overview. Pediatric Nephrology 5, 513-521.

Wanke R, Wolf E, Hermanns W, Folger S, Buchmüller T, Brem G (1992) The GH-transgenic mouse as an experimental model for growth research: clinical and pathological studies. Hormone Research 37, supplement 3, 74-87.

Wanke R, Weis S, Kluge D, Kahnt E, Schenck E, Brem G, Hermanns W (1994) Morphometric evaluation of the pancreas of growth hormone-transgenic mice. Acta Stereologica 13, 3-8.

Wanke R, Wolf E, Brem G, Hermanns W (1995a) Reexpression of the insulin-like growth factor-2 gene in hepatocellular carcinomas from growth hormone transgenic mice. Pathology Research and Practice 191, 190 (Abstract).

Wanke R, Wolf E, Hünemörder S, Erhard M, Brem G, Hermanns W (1995b) Sex-specific stimulation of skin growth by growth hormone in transgenic mice. Experimental and Clinical Endocrinology & Diabetes 103, supplement 1, 62 (Abstract).

Wolf E, Rapp K, Brem G (1991a) Expression of metallothionein-human growth hormone fusion genes in transgenic mice results in disproportionate skeletal gigantism. Growth, Development & Aging 55, 117-127.

Wolf E, Rapp K, Wanke R, Hermanns W, Pirchner F, von Butler-Wemken I, Brem G (1991b) Growth characteristics of metallothionein-human growth hormone transgenic mice as compared to mice selected for high eight-week body weight and unselected controls. II. Skeleton. Growth, Development & Aging 55, 237-248.

Wolf E, Wanke R, Hermanns W, Brem G, Pirchner F, von Butler-Wemken I (1991c) Growth characteristics of metallothionein-human growth hormone transgenic mice as compared to mice selected for high eight-week body weight and unselected controls. I. Body weight gain and external body dimensions. Growth, Development & Aging 55, 225-235.

Wolf E, Kahnt E, Ehrlein J, Hermanns W, Brem G, Wanke R (1993) Effects of long-term elevated serum levels of growth hormone on life expectancy of mice: lessons from transgenic animal models. Mechanisms of Ageing and Development 68, 71-87.

Wolf E, Kramer R, Blum WF, Föll J, Brem G (1994) Consequences of postnatally elevated insulin-like growth factor-II in transgenic mice: endocrine changes and effects on body and organ growth. Endocrinology 135, 1877-1886.

Wolf E, Blum WF, Föll J, Erhard M, Blackburn A, Wanke R, Hermanns W, Brem G (1995a) Regulation of circulating IGFBP-2: new aspects from transgenic mouse models. Experimental and Clinical Endocrinology & Diabetes 103, supplement 1, 95 (Abstract).

Wolf E, Rapp K, Blum WF, Kolb H, Brem G (1995b) Skeletal growth of transgenic mice with elevated levels of circulating insulin-like growth factor-II. Growth Regulation (In Press).

Wolf E, Wanke R, Schenck E, Hermanns W, Brem G (1995c) Effects of growth hormone over-production on grip strength of transgenic mice. European Journal of Endocrinology 133, 735-740.

Wolf E, Brem G (1991) "High-dose hook effect" as a pitfall in quantifying transgene expression in metallothionein-human growth hormone (MT-hGH) transgenic mice. Clinical Chemistry 37, 763-765.

Zambrowicz BP, Harendza CJ, Zimmermann JW, Brinster RL, Palmiter RD (1993) Analysis of the mouse protamine 1 promoter in transgenic mice. Proceedings of the National Academy of Sciences of the USA 90, 5071-5075.

Prospectives of Transgenic Animals for the Pharmaceutical Indudstry

K. Bürki

Summary

Transgenic laboratory animals will be of practical importance for the pharmaceutical industry with regard to several aspects in supporting drug finding and drug development:
(1) by modifying candidate genes, the relevance of potential drug targets can be investigated;
(2) genetically modified animals serve as models for human diseases;
(3) relevant drug receptors are replaced by human (or viral) effector molecules in order to test drug efficacy *in vivo*; and
(4) oncomice, metabolic mutants or animals carrying reporter genes will serve as sensitive indicators for unwanted drug-induced side effects.

For all the above mentioned concepts, relevant examples are available and are being evaluated. However, methodological improvements will allow the establishment of more refined models with a better control of the induced genetic and physiological changes. This may result in a reduction of the overall numbers of animals used for drug development, but also in a reduction of the potentially negative effects of transgenesis on the welfare of the animals.

Introduction

The spectrum of methods to modify the genetic make-up of higher organisms, particularly laboratory rodents, in a stable fashion, is currently being extended by methods of unprecedented precision. Whereas classical transgene constructs are used to insert additional sequences randomly into the genome, today's targeting vectors allow the modification of a selected endogenous gene in a precise and predetermined way. This has opened up new possibilities to re-evaluate a number of fundamental biological questions concerning development and the interaction of cells and organs of higher organisms, in particular to investigate the role of the molecular components of complex *in vivo* systems (Hanahan 1989). The major obstacles in evaluating transgenic models are the usually unreliable expression of many of the randomly-inserted transgenes, and the biological complexity of higher organisms which often obscures the primary effect of the induced genetic alteration (Strohman 1994).

Nevertheless, transgenic models have a powerful potential in many fields of biomedical research, including drug finding and drug development processes. For the pharmaceutical industry, transgenic laboratory animals will therefore be of practical importance with regard to several aspects: (1) Transgenic animals will be designed to gain a better understanding of complex *in vivo* systems, as for example the immune system, the endocrine system or the central nervous system, in the normal and in a pathological state. This will result in the identification of relevant drug targets for the prevention or correction of pathological deviations from normal physiological states. The identification of relevant drug targets is a prerequisite for the design of efficient drug finding programs. (2) Animal models for human diseases will be used to develop and test new therapeutic strategies, as for example gene therapy approaches. (3) Transgenic animals with viral or human drug receptors will replace conventional animals for the testing of drug availability and drug efficacy. (4) Finally, animals with modified metabolic pathways, or animals carrying specific reporter genes, will serve as sensitive indicators of unwanted drug-induced side effects. For all these aspects, transgenic animal models are under investigation.

The present overview will be restricted to selected examples. More extensive reviews have recently been published (Liggitt *et al.* 1992, Bürki and Ledermann 1995). Although of interest for the pharmaceutical industry, the use of transgenic farm animals as producers of therapeutic proteins (Logan 1993), or as organ donors for xenotransplantation (Lu *et al.* 1994) will not be covered. However, some aspects relating to animal welfare will be discussed in a separate paragraph.

Transgenic Animal Models to Support Drug Finding

Many highly successful therapeutic drugs have been selected and developed by entirely empirical approaches. For many of these drugs the molecular drug target, the drug receptor, and the exact mode of action have not been elucidated. However, the identification of relevant drug targets is in many cases mandatory for the identification of agents with higher specificity and less side-effects. Also, the design of high capacity screening systems, based on molecular interactions or on cellular responses, is dependent on an insight into the molecular mechanisms of pathology.

This can be illustrated nicely by discussing progress in modelling Alzheimer's disease in transgenic mice. Alzheimer's disease, a form of senile dementia, is characterized by two kinds of pathological deposits in the brain, called plaques and tangles. Neuritic plaques contain extracellular deposits of β-amyloid, a peptide derived proteolytically from the much larger amyloid precursor protein. Neurofibrillary tangles form inside large pyramidal cells and other cells and consist primarily of abnormal paired helical filaments. These are composed of the microtubule associated protein *tau* in a hyperphosphorylated form. The degree of dementia observed in affected patients appears to correlate better with the extent of neurofibrillary pathology than with the number of plaques (Arriagada *et al.* 1992). The processes leading to the formation of plaques and tangles are not yet understood. Studies in humans are hampered by the inaccessability of the brain. Progress towards an understanding of the pathogenic mechanisms at work in the disease would be supported by the availability of an experimental animal model. Various attempts have been made to generate transgenic mice that exhibit Alzheimer-like pathology, but until recently the results had not been encouraging (Lamb 1995). However, two newer studies indicate that it may be possible to mimic aspects of the human pathology in transgenic mice (Games *et al.* 1995, Götz *et al.* 1995, reviewed by Crowther 1995).

The transgenic mice described by Games *et al.* (1995) express high levels of human mutant amyloid precursor protein. They progressively develop many of the pathological hallmarks of Alzheimer's disease, including numerous extracellular β-amyloid deposits, neuritic plaques, synaptic loss, astrocytosis and microgliosis. However, although many plaques were associated with dystrophic neurites, no neurofibrillary pathology has been described so far.

Changes similar to those that precede the full neurofibrillary pathology in Alzheimer's disease have been observed in a transgenic mouse described by Götz *et al.* 1995. These mice express specifically in their brains the longest human brain *tau* isoform. As in Alzheimer's disease, transgenic human *tau* protein was present in nerve cell bodies, axons and dendrites, and was not restricted to axons as occurs in normal nerve cells. Moreover, the human *tau* was phosphorylated at sites that are hyperphosphorylated in paired helical filaments.

These models, separate or combined, may be useful for identifying some of the steps involved in the pathological cascades of Alzheimer's disease. Relevant steps may be used to develop an *in vitro* high capacity screening system to rapidly screen for potential drugs. The transgenic mice may then serve to test such drugs *in vivo* for their potential to inhibit or reverse the assembly of β-amyloid or the hyperphosphorylation of *tau* protein.

Transgenic Disease Models to Support the Development of New Therapeutic Approaches

Apart from drug-mediated therapies, cell- and gene-mediated therapeutic approaches have been proposed recently or have already been evaluated in patients. Human diseases caused by single gene defects, are among the first to be treated by gene therapy approaches (Morgan and Anderson 1993). Again, an experimental animal model may support the development of strategies of gene delivery and may be indicative of the *in vivo* behaviour of therapeutic gene constructs.

The complexity in generating an experimental animal model for a disease due to a single gene defect is illustrated by the induction of cystic fibrosis (CF) mutations in mice by gene targeting techniques (Clarke *et al.* 1992, Snouwaert *et al.* 1992, Dorin *et al.* 1992). CF is caused by a defective CF transmembrane conductance regulator (*cftr*) gene which encodes a chloride channel. Depending on the targeting construct used, the animals homozygous for the *cftr* disruption display a range of classical symptoms of CF patients. However, in one model a very mild form of cystic fibrosis was observed (Dorin *et al.* 1992), whereas in the other two models very severe forms of the disease were manifested. In fact, these two models had limited utility because most mice died from intestinal obstruction during the first month of life. On the other hand, the surviving mice had mild changes in their lungs and did not show the pathology characteristic of severely affected human CF patients (for review see Smithies 1993). In order to generate a more robust model for the effects of the null CF mutation on the respiratory tract, Zhou *et al.* 1994 expressed human CFTR in the intestine of mutant mice using a transgene under the control of the rat intestinal fatty-acid binding protein gene promoter. These mice survived and showed functional correction of the intestinal pathology. The model may now be suitable to study long-term corrective effects of gene therapy approaches (Hyde *et al.* 1993).

Transgenic Models for Drug Testing

Potential obstacles to the extrapolation of pharmacological data from animals, in particular laboratory rodents, to humans are not only the differences in biochemical pathways, physiological processes or organ systems, but also structural differences at the molecular level. These structural differences may, for example, drastically influence the interaction of a potential antagonist with a receptor of rodent versus human origin. To overcome structural differences, transgenic techniques may be used to introduce drug targets of human structure into rodents, either on wild-type or on a knockout background. In a more sophisticated approach, by using gene targeting strategies, a murine gene might directly be replaced by human sequences (Zou *et al.* 1994).

An example to illustrate this concept is mice transgenic for the human multi-drug resistance gene *MDR*1, encoding the cell surface multi-drug transporter P-glycoprotein. Mice have been created that express human P-glycoprotein on bone marrow cells. These mice become resistant to the leucopenia induced in non-transgenic control animals by chemotherapeutic drugs (Galski *et al.* 1989, Mickisch *et al.* 1991). Bone marrow resistance in the transgenic mice is however reversed significantly by simultaneous administration of known multi-drug transporter inhibitors such as quinidine. These mice should prove valuable as a rapid test system to determine the *in vivo* efficacy of agents with the potential to reverse multi-drug resistance in humans. Rodents with humanized drug receptors may not only speed up the *in vivo* testing of therapeutic agents, but may also, to some extent, replace testing with primates.

Similarly, transgenic mice expressing viral drug targets allow pharmacological studies whilst avoiding the handling of virally contaminated animals (Mehtali *et al.* 1992).

Transgenic Animals to Probe for Drug-induced Side Effects

Toxic, oncogenic or mutagenic effects usually prevent the further development of a drug. Therefore, after testing in rapid *in vitro* screening systems, drug-induced side effects are monitored in laboratory rodents. Subtle side effects may necessitate the use of large numbers of treated animals in order to establish the findings statistically. Transgenic animals sensitized due to an induced genetic predisposition might reduce the test phases as well as the number of animals used. Based on this concept, several transgenic models have been generated and are being evaluated for use in drug testing.

Several transgenic mice harbouring oncogenes, so-called oncomice, have been proposed as sensitive indicator organisms for oncogenic compounds. For example, Breuer *et al.* (1989) have described a transgenic mouse line overexpressing the *pim*-1 oncogene in lymphocytes. These mice are predisposed to develop T-cell lymphomas, but only to the extent that about 10% of the mice develop a lymphoma within 240 days. The mice were tested for their susceptibility to N-ethyl-N-nitrosourea (ENU), a chemical carcinogen. After a single low dose of ENU, nearly all transgenic mice developed T-cell lymphomas within 200 days. These mice are currently being evaluated as sensitive test animals to assess the oncogenic potential of different chemical compounds. As an alternative to oncomice, mice with an induced disruption of a tumour suppressor gene, for example *p53*, may serve as sensitive indicator animals (Harvey *et al.* 1993). Such strains, once validated and generally accepted, may drastically reduce the number of animals necessary for oncogenicity studies.

With regard to testing for genotoxic effects, a number of transgenic animals harbouring reporter constructs flanked by bacteriophage *lambda* sequences have been established. The bacteriophage sequences allow the rapid recovery and recloning of the indicator transgenes, usually bacterial *lacZ* sequences, from any organ of choice of mutagen treated animals. These systems thus allow the fast recovery of *in vivo* induced gene mutations and the direct correlation of mutational mechanisms with, for example, cancer or ageing (for reviews see Gossen and Vijg 1993, Mirsalis *et al.* 1995). Again testing for genotoxic effects will be possible with significantly reduced numbers of animals.

Considerations Related to Animal Welfare

The direct implications of transgenic techniques for animal welfare have recently been discussed by Moore and Mepham, 1995. Whereas the implications for animal welfare before and during the generation of transgenic animals can be controlled and minimized by experienced experimenters, the newly generated transgenic animals as well as their offspring may exhibit reduced welfare. The major factors are (1) direct effects of the introduced genetic alteration, for example in disease models, (2) indirect effects related to the method used, for example insertional mutations after random insertion of transgenes (Meisler 1992), and (3) indirect effects related to the breeding strategy chosen to propagate an induced mutation, for example by enforced inbreeding.

In practice, the occurrence of malformed offspring is very rare due to two biological filters: death *in utero*, and cannibalism by the mother (Overbeek

1994). However, if transgenic animals start to show deviations from normal phenotype or behaviour, a careful analysis will be necessary to determine the exact cause. For a model to be scientifically useful, any indirect effects of the methods or of the breeding strategy used has to be eliminated or at least to be strictly controlled. Whereas this requirement is usually met with regard to accidentally introduced mutations, the importance of the breeding strategy is often neglected. For example, the potential deleterious effect of an insertional mutation can be controlled by breeding transgenes in the hemizygous form. In addition, this breeding strategy has two major advantages: (1) in each generation transgenic animals and its corresponding non-transgenic control littermates are generated; and (2) the transgene can be crossed into a well characterized background.

In most cases, transgenic animals generated by microinjection are of a mixed genetic background, due to the significantly higher efficiency of the microinjection method when using hybrid zygotes or zygotes derived from hybrid parents (Brinster *et al.* 1985). To facilitate further breeding, animals of early offspring generations are often used to cross the inserted transgene to homozygosity. The homozygous line is then propagated by a few individuals only. This will most likely lead to the generation of a new inbred mouse line. Inbreeding of new chromosomal combinations however often results in inbreeding depression (Green 1975), manifested for example by a loss of reproductive fitness. The same considerations are true for the establishment of new lines of mice with targeted mutations. Most embryonic stem cells currently used for gene targeting have been derived from the inbred strain of mice 129, a strain otherwise rarely used in experimental biology. Therefore, many induced targeted mutations are back-crossed into well characterized strains, like C57BL/6 or BALB/c. If back-crossing is stopped after a few generations only, as outlined above, a new combination of chromosomes may eventually be inbred. Moreover, as the effects of genetic traits are variable depending on the genetic background (Nadeu *et al.* 1995), variability is introduced when new substrains are generated.

In particular in pharmacology and toxicology, the availability of animals with a defined, stable genetic background will be mandatory. The availability of embryonic stem cells derived from common inbred strains of mice (Ledermann and Bürki 1991, Noben-Trauth *et al.* 1995) will allow the direct introduction of targeted transgenes or mutations into a well-characterized, defined genetic background and to some extent will replace the strains and strain combinations used presently.

Conclusions

Transgenic animals are valuable tools in many areas of biomedical research. The shortcomings of existing models may be circumvented by the availability of new approaches for an improved spatial and temporal control of transgene expression (Bronson and Smithies 1994, Gossen and Bujard 1992, Shockett 1995), and of the induction of recombination events (Gu *et al.* 1993, Kühn *et al.* 1995). This will lead to an increased demand for refined animal models. Significant numbers of animals will have to be sacrificed to establish new models. However, the better control of induced genetic and physiological changes will lead to a more significant model and, if the experiments are correctly designed, to reduced negative effects on the welfare of the mutant animals. Indirectly, a better understanding of *in vivo* interactions will allow the establishment of improved *in vitro* systems to study molecular mechanisms or to test for adverse drug effects. Sensitive indicator animals designed to test for adverse drug effects *in vivo* will eventually also lead to a significant reduction of the number of animals used in toxicological studies. Animals with humanized drug receptors may help to reduce predictive studies with primates.

References

Arriagada PV, Growden JH, Hedley-White ET, Hyman BT (1992) Neurofibrillary tangles but not senile plaques parallel duration and severity of Alzheimer's disease. Neurology 42, 631-639.

Breuer M, Slebos R, Verbeek S, van Lohuizen M, Wientjens E, Berns A (1989) Very high frequency of lymphoma induction by a chemical carcinogen in pim-1 transgenic mice. Nature 340, 61-63.

Brinster RL, Chen HY, Trumbauer ME, Yagle MK, Palmiter RD (1985) Factors affecting the efficiency of introducing foreign DNA into mice by microinjecting eggs. Proceedings of the Academy of Sciences USA 82, 4438-4442.

Bronson SK, Smithies O (1994) Altering mice by homologous recombination using embryonic stem cells. The Journal of Biological Chemistry, 269, 27155-27158

Bürki K, Ledermann B (1995) Transgenic animals as pharmacological tools. Advances in Drug Research 26, 143-177.

Clarke LL, Grub BR, Gabriel SE, Smithies O, Koller BH, Boucher RC (1992) Defective epithelial chloride transport in a gene-targeted model of cystic fibrosis. Science 257, 1125-1128.

Crowther RA (1995) Steps towards a mouse model of Alzheimer's disease. BioEssays 17, 593-595.

Dorin JR, Dickinson P, Alton EWFW, Smith SN, Geddes DM, Stevenson BJ, Kimber WL, Fleming S, Clarke AR, Hooper ML, Anderson L, Beddington RSP, Porteous DJ (1992) Cystic fibrosis in the mouse by targeted insertional mutagenesis. Nature 359, 211-215.

Galski H, Sullivan M, Willingham MC, Chin KV, Gottesman MM, Pastan I, Merlino GT (1989) Expression of a human multidrug resistance cDNA (MDR1) in the bone marrow of transgenic mice: resistance to daunomycin-induced leukopenia. Molecular and Cellular Biology 9, 4357-4363.

Games D, Adams D, Alessandrini R, Barbour R, Berthelette P, Blackwell C, Carr T, Clemens J, Donaldson T, Gillespie F, Guido T, Hagoplan S, Johnson-Wood K, Khan K, Lee M, Leibowitz P, Lieberburg I, Little S, Masliah E, McConlogue L, Montoya-Zavala M, Mucke L, Paganini L, Penniman E, Power M, Schenk D, Seubert P, Snyder B, Soriano F, Tan H, Vitale J, Wadsworth S, Wolozin B, Zhao J (1995) Alzheimer-type neuropathology in transgenic mice overexpressing V717F β-amyloid precursor protein. Nature 373, 523-527.

Gossen M, Bujard H (1992) Tight control of gene expression in mammalian cells by tetracycline-responsive promoters. Proceedings of the National Academy of Sciences USA 89, 5547-5551.

Gossen J, Vijg J (1993) Transgenic mice as model systems for studying gene mutations in vivo. Trends in Genetics 9, 27-31.

Götz J, Probst A, Spillantini MG, Schäfer T, Jakes R, Bürki K, Goedert M (1995) Somatodendritic localization and hyperphosphorylation of tau protein in transgenic mice expressing the longest human brain tau isoform. The EMBO Journal 14, 1304-1313.

Green EL (1975) Breeding systems. In: Biology of the Laboratory Mouse (Green EL, ed). New York: Dover, pp 11-22.

Hanahan D (1989) Transgenic mice as probes into complex systems. Science 246, 1265-1275.

Harvey M, McArthur MJ, Montgomery CA, Butel JS, Bradley A, Donehower LA (1993) Spontaneous and carcinogen-induced tumorigenesis in p53-deficient mice. Nature Genetics 5, 225-229.

Hyde SC, Gill DR, Higgins CF, Trezise AEO, MacVinish LJ, Cuthbert AW, Ratcliff R, Evans MJ, Colledge WH (1993) Correction of the ion transport defect in cystic fibrosis transgenic mice by gene therapy. Nature 362, 250-255.

Kühn R, Schwenk F, Aguet M, Rajewski K (1995) Inducible gene targeting in mice. Science 269, 1427-1429.

Lamb BT (1995) Making models for Alzheimer's disease. Nature Genetics 9, 4-6.

Ledermann B, Bürki K (1991) Establishment of a germ-line competent C57BL/6 embryonic stem cell line. Experimental Cell Research 197, 254-258.

Liggitt HD, Omiecinski CJ, Quaife CJ (1992) Potential applications of transgenic animals in drug development. Progress in Drug Metabolism 13, 1-33.

Logan JS (1993) Transgenic animals: beyond 'funny milk'. Current Biology 4, 591-595.

Lu CY, Khair-El-Din TA, Dawidson IA, Butler TM, Brasky KM, Vazquez MA, Sicher SC (1994) Xenotransplantation. The FASEB Journal 8, 1122-1130.

Mehtali M, Munschy M, Ali-Hadji D, Kieny MP (1992) A novel transgenic mouse model for the in vivo evaluation of anti-human immunodeficiency virus type 1 drugs. AIDS Research and Human Retroviruses 8, 1959-1965.

Meisler MH (1992) Insertional mutation of 'classical' and novel genes in transgenic mice. Trends in Genetics 8, 341-345.

Mickisch GH, Merlino GT, Galski H, Gottesman MM, Pastan I (1991) Transgenic mice that express the multidrug-resistance gene in bone marrow enable a rapid identification of agents that reverse drug resistance. Proceedings of the National Academy of Sciences USA 88, 547-551.

Mirsalis JC, Monforte JA, Winegar RA (1995) Transgenic animals for the detection of in vivo mutations. Annual Review of Pharmacology and Toxicology 35, 145-164.

Moore CJ and Mepham TB (1995) Transgenesis and animal welfare. ATLA 23, 380-397.

Morgan RA and Anderson WF (1993) Human gene therapy. Annual Review of Biochemistry 62, 191-217.

Nadeu JH, Arbuckle LD, Skamene E (1995) Genetic dissection of inflammatory responses. Journal of Inflammation 45, 27-48.

Noben-Trauth N, Köhler G, Bürki K, Ledermann B (1995) Efficient targeting of the IL-4 gene in a BALB/c embryonic stem cell line. Transgenic Research (in press).

Overbeek P (1994) Factors affecting transgenic animal production. In: Transgenic Animal Technology: A Laboratory Handbook (Pinkert CA, ed). San Diego: Academic Press, pp 69-114.

Shockett P, Difilippantonio M, Hellman N, Schatz DG (1995) A modified tetracycline-regulated system provides autoregulatory, inducible gene expression in cultured cells and transgenic mice. Proceedings of the National Academy of Sciences USA 92, 6522-6526.

Smithies O (1993) Animal models of human genetic diseases. Trends in Genetics 9, 112-116.

Snouwaert JN, Brigman KK, Latour AM, Malouf NN, Boucher RC, Smithies O, Koller BH (1992) An animal model for cystic fibrosis made by gene targeting. Science 257, 1083-1088.

Strohman R (1994) Epigenesis: The missing beat in biotechnology? Biotechnology 12, 156-164.

Zhou L, Dey CR, Wert SE, DuVall MD, Frizzell RA, Whitsett JA (1994) Correction of lethal intestinal defect in a mouse model of cystic fibrosis by human CFTR. Science 266, 1705-1708.

Zou YR, Müller W, Gu H, Rajewski K (1994) CRE-loxP-mediated gene replacement: a mouse strain producing humanized antibodies. Current Biology 4, 1099-1103.

Assessing the Welfare of Transgenic Animals

D. M. Broom

Summary

Techniques for the scientific assessment of the welfare of animals have developed rapidly in recent years and many of these can be applied to genetically modified animals. Each modified strain should be compared with unmodified strains using measures of physiology, behaviour, anatomy, immune system function, pathological change, growth, reproduction and longevity. Using a wide range of measurements, any increased levels of pain, fear or distress should be revealed. There may also be indications of the extent to which the welfare of such individuals can be good.

A potential problem in using some welfare assessment techniques is that the animal may be affected by the genetic modification in a way which alters the aspect of its biology which is being measured. For example, if a preference is tested but the relevant sensory functioning has been genetically altered or if an adrenal response is to be measured and adrenal functioning has been changed in the modified animals, then the measurement procedure would be invalid. Such problems must be considered wherever the welfare of transgenic animals is to be assessed.

Every person who works with transgenic animals should be aware of how to assess their welfare and should act so as to avoid or minimise poor welfare. When a transgenic animal has been developed, details of any effects of the genetic manipulation on the welfare of the animal should be part of the specification available for potential users. Whilst there is some legislation concerning the welfare of animals which are part of experiments, in most countries the only legislation relevant to their welfare after this is of a general nature, for example that concerning cruelty to animals. Such legislation is not adequate for transgenic animals. The legislation should stipulate that no genetically modified animal should be permitted to be used commercially until comprehensive studies of the welfare of the animal have been carried out during two generations and continuing for maximum commercial life. The decision as to whether the use of the modified animal is permitted should depend upon whether there is a net benefit for the welfare of all animals including humans. A commercial profit is not sufficient justification for modifying an animal in such a way that its welfare is poor.

Introduction

If animals are to be produced as a consequence of transgenic procedures, two important questions need to be answered. These are - whether or not there are positive or negative effects on welfare and the magnitude of those effects? Hence it is essential to use a definition of welfare which allows scientific measurement. The welfare of an animal is its state as regards its attempts to cope with its environment (Broom 1986). This state refers to the amount of difficulty which the individual has in trying to cope with its environment and the extent to which it is failing to cope. When it fails to cope, or seems likely to do so, it is said to be stressed. The state of the animal includes the feelings of the individual, which may be good feelings or suffering (Broom 1995b). In order to be a useful scientific concept, we must be able to think of welfare as varying over a range from very good to very poor.

There is a rather small range of measures which give us information about welfare at the good end of the scale (Table 1), but a much longer list of measures which can tell us about how poor the welfare of the animal is (Table 2). All of these measures and the concept of welfare are explained in detail by Broom (1991) and Broom and Johnson (1993).

Table 1. Measures of good welfare

Variety of normal behaviours shown
Extent to which strongly preferred behaviours can be shown
Physiological indicators of pleasure
Behavioural indicators of pleasure

(from Broom and Johnson, 1993)

Table 2. Measures of poor welfare

Reduced life expectancy
Reduced ability to grow or breed
Body damage
Disease
Immunosuppression
Physiological attempts to cope
Behavioural attempts to cope
Behavioural pathology
Self narcotization
Extent of behavioural aversion shown
Extent of suppression of normal behaviour
Extent to which normal physiological processes and anatomical development are prevented.

(from Broom and Johnson, 1993)

Measurements of animal welfare should be made in an objective, scientific way. Once they are made, moral judgements concerning what is acceptable can be made more easily. However, the process of scientific evaluation should be kept separate from the moral judgement.

Conventional Breeding and Welfare

Conventional breeding methods can change animals in such a way that they have more difficulty in coping or are more likely to fail to cope (Broom 1994, 1995a). One example of such an effect is the sensory, neurological or orthopaedic defect found commonly in certain breeds of dog. Others are the effects of the genes promoting obesity in mice, double muscling linked to parturition problems in cattle and many examples of selection promoting fast growth and large muscles in farm animals. Modern strains of pigs have relatively larger muscle blocks, more anaerobic fibres and smaller hearts than have the ancestral strains (Dämmrich 1987). They are more likely to die or to become distressed during any activity. Modern broiler strains grow to a weight of 2 - 2.5 kg in 37 days as compared with 12 weeks thirty years ago. Indeed, the maturation age is decreasing by one day per year at present. Their muscles and guts grow very fast, but the skeleton and cardio-vascular system do not. Hence, many of the birds have leg problems, such as tibial dyschondro-plasia or femoral head necrosis, or cardiovascular malfunction which may cause ascites.

It is clear that the welfare of meat producing animals, which are growing too fast for their legs and heart, is becoming poorer and poorer because of this genetic selection and that the continuation of this trend is morally wrong. The competitive nature of the industry makes it difficult for individual producers to take action to reverse the trend, but many of them are now breaking the cruelty laws. This point is made here because there is pressure on those concerned with genetic engineering to make such animals grow even faster.

The Welfare of Transgenic Animals

In this section it is emphasised that transgenesis can result in better welfare, in no change from the average for unmodified animals, or in poorer welfare. Some of the points concerning welfare assessment are explained in more detail later in this paper.

Some genetic manipulations can be beneficial to the modified animals. If genes conferring disease resistance are inserted into the genome of an individual, for example in the work on avian leucosis virus resistance, then the welfare of the modified individual is better than that of the unmodified individual. If the animal can cope with disease challenge better then its welfare is slightly improved for most of the time and very much improved in the circumstance where disease challenge occurs.

When the transgenic animal is modified so that it can produce a novel protein in its blood or milk, there may be no effect at all on its welfare. However there could be some adverse effect and the predictability of that effect will vary according to the precision of the transgenesis procedure. Gene transfer by introducing embryonic stem cells into a blastocyst are more predictable in their effects than the introduction of genetic material by microinjection.

The production of disease susceptible animals by transgenesis, so that the animals can be used in medical research, will result in poorer welfare whenever the gene is expressed. The extent of the poor welfare will differ considerably according to the level of expression and the disease state.

If, as discussed in one of the preceding chapters, the animals produced as a result of transgenesis were modified in a way which increased their growth rate, or the growth of a particular organ, or differential growth in such a way that an already productive genetic strain was made even more productive, there is a serious risk that the welfare of the animals would be worse as a direct consequence of the manipulation. Those carrying out such work should consider whether the animals are already close to some biological limit on adaptability before proceeding.

Genetic manipulation could affect: - sensory functioning; the structure of bones or muscles; hormone production; detoxification ability; neural functioning etc. The question which must be considered is not whether or not there is a change but whether there is a change which affects the animal's welfare. In some cases, any effects of the genetic modification on the welfare of other individuals must be considered.

In a study of the effects of transgenesis on welfare, control animals which have not been modified should also be used. A wide range of measures of welfare are necessary because the actual effects on the individual will seldom be known and also because species and individuals vary, both in the methods which they use to try to cope with adversity and in the measurable signs of failure to cope. A simple welfare indicator could show that welfare is poor, but absence of an effect on one indicator of poor welfare does not mean that the welfare is good. For example, if the major effect of a manipulation was a behavioural abnormality or an increase in disease susceptibility but only growth rate was measured, then a spurious result could be obtained. The choice of measurements should include the main methods of assessing poor welfare which are mentioned here. Often it will be obvious from a preliminary study of morphology, or a clinical examination, which measurements of function or of pathology will be most relevant.

The effects of genetic manipulation may not be apparent at all stages of life, so the animal must be studied at different stages including the oldest age likely to be reached during usage. Some effects may be evident in the second generation but not in the first, so modified animals should be studied for two generations.

Measures of Welfare

Preference Studies

As listed in Table 1, an important technique in welfare research is the measurement of the strength of animal preferences. Studies of positive preferences involve choice tests, often with some operant technique being used to indicate how hard the individual will work to obtain a particular resource or have the opportunity to carry out a certain behaviour (Dawkins 1983, Arey 1992, Manser *et al.* 1995). A possible problem which must be considered when using such methods is that the sensory or motor ability of the animal might be altered by the transgenesis. Positive preferences could on occasion give ambiguous results, but in general it would be expected that what is important to normal animals would also be important to transgenic animals. Studies of aversion and its strength would be of value in studies of transgenic animals. If, for example, the modified animal were changed so that bright light was aversive, the extent of the aversion could be measured in studies of actual movement away from light, of reluctance to be moved towards a well lit place or of some specific task which had to be performed in order to avoid the onset of bright light.

Reproductive Success

Some zoo animals cannot breed when potential breeding partners are present, because of an inadequacy in their environment. The welfare of these animals is less good than that of animals which can breed. Inability to reproduce would be an indicator of poor welfare in transgenic animals.

Growth, Weight Loss, Mortality and Life Expectancy

If control animals can grow or maintain weight in a given situation but modified animals fail to grow or lose weight, this would indicate poorer welfare in the latter. Abnormal weight gain could also indicate a problem. It is important to use a biologically relevant control in such studies. An animal could be losing weight because it is lactating or is a reproductively active male, like a red deer in rut. On the other hand, an animal which is in the pre-hibernation condition could put on a great deal of weight.

Measures of mortality rates have long been used in studies of the effects of housing conditions or management methods on animal welfare. As Hurnick and Lehman (1988) have pointed out, a housing condition, management method or treatment which resulted in the animal having a lower life expectancy indicates poorer welfare with that condition or treatment. Indeed a human who died early because of some form of self abuse or an energetic life style would be considered to have been under greater stress than a similar but longer lived person. Other examples include cetaceans which die early in poor zoo conditions and dairy cows which do not live as long under the very high production conditions of recent years as they did when their metabolic pace of life was lower (Agger 1983, Broom 1993).

Physiological Measures

Aspects of normal physiological functioning, for instance of the kidneys, could be affected in some genetically modified animals. Some of the abnormalities would be detected by clinical examination but others require specific tests to be carried out for detection.

Several physiological measurements are of value in assessing the extent to which emergency responses have been used by an individual. When there is a short-term problem, the individual may increase its heart rate and adrenal activity. Transgenic animals could be tested in situations in which control animals would show a known mean level of physiological response in order to ascertain whether or not those situations caused them more problems. It might also be useful to investigate longer term usage of adrenal cortex responses by means of dexamethasone and adrenocorticotrophic hormone challenge tests (Dantzer and Mormède 1983, Mendl *et al.* 1992).

A further method of coping with adversity is to use endogenous opioids in the brain to self narcotise. The welfare of individuals which have to do this is less good than that of those who do not. The measurement of levels of plasma opioids appears to give little information about this coping method and the experimental use of opioid antagonists is difficult to interpret. However, studies of opioid receptor density may prove useful in welfare assessment (Zanella *et al.* 1991).

Measures of Immune System, Disease and Injury

When animals show substantial adrenal cortex responses, this is often associated with some degree of immunosuppression (Kelley 1985, Siegel 1987). There are also other mechanisms by which difficult conditions lead to impairment of immune system function. Measurements of immunosuppression include antigen challenge tests, lymphocyte stimulation tests *in vivo*, lymphocyte proliferation tests *in vitro* and specific tests of natural killer cell or macrophage efficacy (Broom and Johnson 1993). If a genetically modified animal had less efficient immunological defences than an unmodified control, then it would be coping less well with its environment so its welfare would be less good. Disease always indicates some effect on welfare, so if that animal was also diseased and suffering then its welfare would be considerably worse. One of the first steps in assessing the welfare of a modified animal is to carry out a thorough clinical examination.

Injury also means poor welfare, the extent depending on the magnitude of the injury and the amount of associated suffering. A predisposition to injury because of weakness of some kind also indicates reduced ability to cope with the environment and hence poor welfare. Hens in battery cages (Knowles and Broom 1990, Norgaard-Nielsen 1990) and sows in stalls (Marchant and Broom 1994, 1996) have weak bones because they get insufficient exercise. If a transgenic animal had thin skin, weak bones or some other effect which predisposed the individuals to injury, then its welfare would be less good than that of controls.

Behavioural Measures

Abnormalities of behaviour are often the easiest indications of poor welfare to recognise and are an integral part of a proper clinical examination. However, careful behaviour recording is also important in welfare assessment and no attempt to assess welfare would be complete unless such recording were carried out. In order to recognise problems in carrying out normal movements, the observer must first establish which movements occur and with what frequency in normal individuals. When Andreae and Smidt (1982) wanted to assess the extent of abnormality of standing and lying movements occurring in young bulls kept on slippery slats they compared these movements with those of bulls on non-slippery floors. In a study of the extent of walking abnormalities in broiler chickens (Kestin *et al.* 1994) classified locomotor ability by its difference from normality and reported that the majority of birds had some locomotor problem before they reached slaughter age.

In studies of the effects of inadequate housing conditions where the animal has insufficient control over important events in its life, stereotypies are sometimes shown. These repeated, relatively invariant movements with no obvious function are readily recognised. Examples are route-tracing in zoo animals, water spout circling in laboratory rodents and crib-biting in horses. Other abnormalities of behaviour include self mutilation, excessive aggression, unresponsiveness, and attention to localised sources of irritation or pain. For further details see Broom and Johnson (1993).

Progress in Welfare Assessment of Transgenic Animals

Some clinical examination will have been made of most transgenic animals and in the more extreme cases where welfare is obviously poor, the experimental study will have been terminated. However, no comprehensive study of the welfare of a transgenic animal has been published. This represents a serious failing on the part of researchers, administrators and governments who have allowed developments to proceed to the point where some of these animals are being used commercially or in medical research. The results of studies of the welfare of the animal should be put in the specification of the animal prepared for subsequent users.

Legislation Required

During experimentation on transgenesis, there is legislation about animal experimentation in the European Union which requires that some account should be taken of the welfare of the animal. Research workers need to consider the welfare of the animal carefully and should be able to justify all of their actions to a member of the general public.

It will not be adequate to depend upon the moral consciences of those who use transgenic animals and specific legislation is needed concerning testing before usage. There is EU legislation relating to human health and preservation of the environment. There should also be legislation requiring that no genetically modified animals should be used commercially unless their welfare has been assessed using an adequate range of measures at suitable intervals throughout life and on through the next generation. If there is a net benefit for the welfare of animals, including humans, then the genetic manipulation should be permitted. This is a stricter criterion than just to say that any harm to the animal must be weighed against any benefit because this latter criterion could allow severe effects solely for financial gain. Modifications of animals which are carried out for commercial purposes only but which result in poor welfare, should not be permitted. There is legislation stating that genetically modified animals cannot be used unless specific permission is given in the Netherlands. The EU and other countries should be following that lead. If such action does not occur quickly it will become more difficult as economic pressures build up.

References

Agger JF (1983) Production disease and mortality in dairy cows. Analysis of records from disposal plants from 1969-1982. In: Proceedings of the 5th International Conference on Production Diseases in Farm Animals. Uppsala, pp 308-311.

Andreae U, Smidt D (1982) Behavioural alterations in young cattle on slatted floors. In: Disturbed Behaviour in Farm Animals (Bessei, ed). Stuttgart, Eugen Ulmer.

Arey DS (1992) Straw and food as reinforcers for prepartal sows. Applied Animal Behaviour Science 33, 217-26.

Broom DM (1986) Indicators of poor welfare. British Veterinary Journal 142, 524-526.

Broom DM (1991) Animal welfare: concepts and measurement. Journal of Animal Science 69, 4167-4175.

Broom DM (1993) A usable definition of animal welfare. Journal of Agricultural and Environ-mental Ethics 6, supplement 2, 15-25.

Broom DM (1994) The effects of production efficiency on animal welfare. In: Biological basis of sustainable animal production. Proceedings of the 4th Zodiac Symposium EAAP, Publication 67 (Huisman EA, Osse JWM, Van der Heide D, Tamminga S, Tolkmap BL, Schouten WGP, Hollingsworth CE, Van Winkel GL, eds). Wageningen: Wageningen Pers, 201-210.

Broom DM (1995a) Measuring the effects of management methods, systems, high production efficiency and biotechnology on farm animal welfare. In: Issues in Agricultural Bioethics, (Mepham TB, Tucker GA, Wiseman J, eds) Nottingham: Nottingham University Press, pp 319-334.

Broom DM (1995b) Animal welfare defined in terms of attempts to cope with the environment. Acta: Agriculturea Scandinavica Section A, Animal Science 45 (in press).

Broom DM, Johnson KG (1993) Stress and Animal Welfare. London: Chapman and Hall. 211 pp.

Dämmrich K (1987) Organ change and damage during stress - morphological diagnosis. In: Biology of Stress in Farm Animals: an Integrated Approach (Wiepkema PR, Van Adrichem PWM, eds). Dordrecht: Martinus Nijhoff, pp 71-81.

Dantzer R Mormède P (1983) Stress in farm animals: a need for re-evaluation. Journal of Animal Science 57, 6-18.

Dawkins M (1983) Battery hens name their price: consumer demand theory and the measurement of animal needs. Animal Behaviour 31, 1195-1205.

Hurnik JF, Lehman H (1988) Ethics and farm animal welfare. Journal of Agricultural Ethics 1, 305-18.

Kelley KW (1985) Immunological consequences of changing environmental stimuli. In: Animal Stress (Moberg GP, ed), Betheseda MD: American Physiological Association, pp 193-223.

Kestin SC, Adams SJM, Gregory NG (1994) Leg weakness in broiler chickens, a review of studies using gait scoring. In: Proceedings of the 9th European Poultry Conference, Glasgow, 7-12th August 1994. Volume II, WPSA, pp 203-206.

Knowles TG, Broom DM (1990) Limb bone strength and movement in laying hens from different housing systems. Veterinary Record 126, 354-356.

Manser CE, Morris TH, Broom DM (1995) An investigation into the effects of solid or grid cage flooring on the welfare of laboratory rats. Laboratory Animals 29, 353-363.

Marchant JN, Broom DM (1994) Effects of housing system on movement and leg strength in sows. Abstract in: Applied Animal Behaviour Science 41, 275-275.

Marchant JN, Broom DM (1996) Effects of dry sow housing conditions on muscle weight and bone strength. Animal Science 62 (in press).

Mendl M, Zanella AJ, Broom DM (1992) Physiological and reproductive correlates of behavioural strategies in female domestic pigs. Animal Behaviour 44, 1107-1121.

Norgaard-Nielsen G (1990) Bone strength of laying hens kept in an alternative system, compared with hens in cages and on deep litter. British Poultry Science 31, 81-89.

Siegel HS (1987) Effects of behavioural and physical stressors on immune responses. In: Biology of Stress in Farm Animals (Wiepkema PR, Van Adrichem PWM, eds), Current Topics in Veterinary Medicine and Animal Science, Dordrecht: Martinus Nijhoff, pp 39-54.

Zanella AJ, Broom DM, Hunter JC 1991. Changes in opioid receptors of sows in relation to housing, inactivity and stereotypies. In: Applied Animal Behaviour: Past, Present and Future (Appleby MC, Horrell RI, Petherick JC, Rutter SM, eds). Potters Bar: Universities Federation for Animal Welfare, pp 140-141.

Production of Transgenic Animals: Practical Problems and Welfare Aspects

P. Costa

Summary

Transgenic technology has been used to produce animals as suitable *in vivo* systems to study human disease, mammalian gene expression and regulation, as well as finding applications in zootechnology, pharmacology and medicine. The past few years have seen a rapid proliferation in the creation of genetically engineered mice as animal models of human diseases, representing an unique and valuable genetic resource.

However genetic manipulation can alter a balanced genotype and produce unpredictable effects.

In order to define, standardise and validate the new transgenic models, it is fundamental to avoid environmental contamination, and to provide special care in terms of space, health management and welfare assessment. Particularly for rodents, the facility in which the animals are kept must be centralised, physically contained and microbiologically controlled. All the animals must be microbiologically defined (specified pathogen free, SPF). Active environmental control, health management, and medical surveillance of personnel must be performed on a constant basis.

Newly generated animal models should be observed following simple but essential criteria. These observations should also be quantifiable and reproducible. Following the ontogenesis of behaviour is a good strategy to recognise and assess deficits before these simple responses are integrated by homeostatic and adaptive mechanisms.

Screening conducted from birth will allow any innate deficits to be identified, estimated and quantified. This screening would also provide the basis for modulating the environment, housing and caging conditions, as well as producing useful information which to date is not included in most scientific publications.

Introduction

Genetic engineering and recombinant DNA techniques are exceptionally powerful tools for progress in biomedical research (Iannaccone *et al.* 1993), and the number of scientific publications on transgenic animals has grown exponentially over the past ten years.

A tremendous amount of work has been done, especially to understand better the mechanisms of gene interaction and integration, to solve the problems of DNA recombination and unpredictable effects in the host, and to refine and develop increasingly specific techniques (Capecchi 1989, Isola *et al.*1991, Melton 1994, Barlow *et al.* 1994, Camper *et al.* 1995). Animal models created through genetic modification have opened new scientific horizons; our ability to predict the impact of these technologies is limited. The most rational way to solve these problems is to identify the general issues to be solved, to assess the risks involved and then to define new standards on the basis of our previously acquired knowledge.

In this contribution some emphasis will be put on the following aspects:
- Environmental control;
- Standardization and health control;
- Welfare assessment.

Environmental Control

Lack of physical containment might lead to transgenic animals mixing with animals of wild populations, which could in the long term, provoke negative effects on the natural biological ecosystem. Species that present particular risks are those which can survive in a natural environment without the assistance of humans and whose phenotypes are competitive to conspecific wild types (Bruggemann 1993). It is difficult to predict what the effects on the environment would be.

Clearly, keeping these transgenics in defined microbiological conditions will make them more dependent on a human controlled environment, which considerably lowers the risks mentioned above. However, the only totally safe way to address this issue is to also house the animals in a physically contained environment, equipped with efficient security systems and regular assessment of the animal's phenotypes.

Standardization and Health Control

Transgenic animals often have a great value because of their unique genetic properties. It is therefore essential to ensure that these animals are free from undesirable infectious diseases (specified pathogen free, SPF). The only way to ensure this is by housing them in a barrier facility.

A biologically and controlled environment which can be defined microbiologically is fundamental to breeding and to working with specified pathogen free animals. The microbiological standardization of the animal corresponds in this case to two fundamental requirements: control of the model, and its standardization and validation.

Housing of transgenic animals should be centralised and structured technologically as a "level 2" biological barrier (Ruys 1991). A transgenic unit must also be designed and conceived as a research unit where different specialists have to interact and work and as a laboratory where everything is planned and laid out in the most rational way possible in terms of space, equipment, human and animal resources. It must be organized accordingly to meet research requirements, while still guaranteeing the quality and microbiological status of the animals housed.

Standard operating procedures (SOPs) and an internal health programme must be followed because:

a) each transgenic animal produced is a potential new model from which we expect specific information;
b) all the variables that possibly could bias the data should be eliminated in order to reliably define a model (standardization gives the true phenotype expression of the gene inserted or eliminated);
c) we need to be certain that any pathologies developed by the animals are due to the genetic manipulation, and not to extrinsic factors.

In order to maintain the SPF status, an internal animal health surveillance programme must be set up to monitor for the presence of pathogens. This includes the use of sentinel animals and the screening of animals purchased from vendors, for instance according to the recommendations of the FELASA working group on animal health (FELASA 1994).

In order to safeguard the pathogen free status of the animals, it is advisable to maintain separate protocols of quarantine for certified SPF animals supplied by breeders and for non-certified SPF animals.

The standard quarantine protocol for certified SPF animals should include verification of the health certificate provided by supplier; examination and observation of the animal on receipt; and a protocol for random screening. The quarantine protocol for non-certified animals should include the receipt of the animals in a "Reception Room" near, but outside the animal house; embryo transfer or Cesarian rederivation using a SPF female; housing the foster mother in a further quarantine area with ventilated hepa filter "in-out" cabinet; and a complete microbiological check of mother and pups at the age of two months.

Intramural health management also includes a medical surveillance programme for all persons involved in animal handling and the microbiological control of the environment and (biological) materials.

Quality control of the animals and standardization of procedures not only contributes to the quality of research but also to reduction of the number of animals used, to refinement of procedures and to a responsible use of animals.

It is clear that the operations in a transgenic animal unit are complex in terms of space requirements, breeding programmes, animal resources and experimental procedures. Also, organizing the flow of traffic for both materials and people, maintenance of the buildings and defined environmental conditions, or the variety of technologies to be used may generate problems. Accomplishing high standards and efficiency in all these areas requires the comprehension and the cooperation of researchers, animal care personnel and line management. So another fundamental issue in the management of a transgenic unit is to ensure ongoing education and training for all those involved with animals.

An education programme carried out by competent personnel can motivate and increase the awareness of personnel as to the high value of the animals housed, and the importance of a welfare assessment programme.

Welfare Assessment

Having described some general standards for establishing and maintaining a transgenic animal programme, I would like to consider the welfare aspects of the animals *per se*.

Animal welfare is defined as an animal's state as regards its attempts to cope with its environment (Broom 1991) and "welfare can generally be visualized as a continuum from good to bad, with well-being considered as the good part of the scale" as stated by Baxter (1983).

We are responsible for the housing of our animals, and it is our duty to keep the animal in such a way that it can be placed on the well-being end of the scale. In the case of newly generated animals, this can be done only by monitoring the animals as they develop, and by establishing what are normal development criteria, and hence provide the right conditions for the well-being of that particular animal.

Interfering with the genotype by inserting or removing fragments of DNA may result in a drastic alteration of the animal's normal genetic homeostasis and this can be manifested in the behaviour and well-being of the animals in unpredictable ways (Pursel and Rexroad 1993, Ting *et al.* 1994, Cockayne *et al.* 1994).

Newly generated animal models should be monitored by using simple, but essential criteria. The observations should be quantifiable and reproducible in order to conduct longitudinal studies in homo- and heterozygous mutant animals and in wild type animals.

Objective indicators of welfare are life span, reproductive capability, growth rate, susceptibility to disease and behavioural alterations (Broom 1991). In particular, the ontogenesis of behaviour might be considered as a sensitive indicator. The use of neurobehavioural tests can be seen as an important tool in the welfare assessment of a genetically modified animal. Screening behaviour from birth will allow eventual deficits to be identified, estimated and quantified. The results of these screenings can be used for modifying the environment, housing and caging conditions. This type of information is usually not included in scientific publications.

For this approach, neurobehavioural tests normally used for the safety assessment of drugs and chemicals can be used. Neurobehavioural tests are used in toxicology studies, and in particular developmental tests are used to study and describe the functional and behavioural changes in offspring resulting from intrauterine insult (behavioural teratology) (Spyker 1975, Rodier *et al.* 1979, Costa *et al.* 1985). For the mouse, modified versions of the tests developed by Fox (1965) and by Altman and Sudarschan (1975) can be used. These tests will reveal information on the ontogenesis of the nervous system. The functions that can be tested for are growth and physical development, neuromotor development, and sensory functions.

All observations are non-invasive and are performed with apparently healthy pups from 1 to 17 days old.

Description of Tests

At birth the litters are weighed and pups are observed for any malformation or special trait. After having been weighed individually and sexed, they are numbered with non-toxic ink and subjected to the battery of tests. These tests are interrupted at the 17th day of life.

The response to each of the tests may be 0 (behaviour or response is absent); 1 (signs of primitive response); 2 (clear but not mature response) or 3 (mature and complete response).

The tests are as follows;

- *Righting* (labyrinth/postural response)

  The animal is put on its back with its limbs pointing upwards. The "mature" response (6-8 days old) is to immediately turn over and stand on its

Figure 1: ***Righting***: *(labyrinth and postural response)*
The animal is put on its back with its limbs pointing upwards. The "mature" response (6-8 days old) is to immediately turn over and stand on its four limbs.

four limbs (Figure 1). The maximum score (3) is given if the animal turns over as soon as it is placed backwards on the surface.

- *Acceleration righting* (statokinetic labyrinth response)
The animal is held by its four limbs and head downwards. It is dropped from about 20 cm. above the surface. The response is mature when the mouse lands perfectly the right way up on its limbs. This normally occurs on day 11 after birth.
- *Grasping*
When the palm area of one of the forelimbs is stimulated with a thin piece of wire, the animal immediately grasps it. In this case, the strength of response also has to be considered in the assessment. This reflex disappears with the development of nervous system.
- *Geotaxic response* (labyrinth/postural reaction)
The mouse is placed on a surface inclined at 45° with its head facing downwards. The mature response is when the animal immediately turns 180° and starts to climb upwards (6-7 days old).
- *Cliff drop aversion* (somato sensory function)
The mouse is placed on the edge of a surface with its forelimbs partially on and partially off this surface. The mature response is that the animal quickly turns its head and forelimbs and avoids dropping.
- *Walking* (locomotion)
This term includes the complete development of locomotion which in rodents has 3 stages:
  - pivoting: moving around with the help of the head and forelimbs but not using the hindlimbs (stage 1)
  - crawling: moving on all four limbs, dragging the belly on the surface (stage 2)
  - walking: mature locomotion, with the body supported completely by the four limbs (stage 3).

  Knowledge of this pattern of development allows to test both for locomotive coordination and for the muscle strength required to support the body.
- *Vibrissa placing* (somato sensory function)
When still blind, the pup is lifted by its tail and its vibrissa are stimulated with a piece of paper. The mature response is to move the head around and to move forelimbs towards the stimulus (all or nothing response).
- *Auditory startle* (sensory function)
The animal jumps at a sudden auditory stimulus.
- *Bar holding* (muscular strength)
The animal is put on a bar. The mature response is when the animal holds the bar with all four limbs with its tail wound round the bar.

- *Crossed extensor* (spinally-mediated reflex)
  Pinching the foot of one hindlimb causes flexion of the stimulated limb and extension of the opposite limb (all or nothing response).

Besides neurobehavioural tests, a screening programme may also include morphological aspects, such as the age of onset of incisor eruption, eyes open, ears open and hair growth.

The above screening programme is presently carried out in our laboratories at I.R.B.M. on newly generated transgenic mice. The screening has allowed us to identify certain deficits in weight development in five out of ten different lines of transgenic animals. These have been a reason to modify our standard housing and breeding protocols in order to compensate for these deficiencies (for instance delaying the time of separation from the mother, improve the accessibility for food and water). With such simple precautions, the transgenic animal can often reach sexual maturity and develop a behaviour that will lead to normal reproduction.

Other neurobehavioural tests should be performed on adult mice to assess other aspects of behaviour, such as learning ability (Paylor *et al.* 1994), locomotion activity and exploration in an unfamiliar environment, which serves as an index of emotional state (Hilakivi-Clarke 1994, Gerlai *et al.* 1994).

Conclusions

Transgenic animals are extremely important for progress in biomedical research. However, their specific characteristics also require that minimum standards are developed to guarantee the welfare of animals and minimize the environmental risks. The system of housing animals in centralised facilities to achieve physical and microbiological containment protects the environment and safeguards the unique genetic properties of the model.

The standardization of environmental conditions and a correct animal care programme aimed at eliminating all exogenous stress factors that could bias the normal development of the animal, allow us to assess welfare problems objectively in newly generated animals, as any alterations of the health status or behaviour can then be attributed to the genetic modifications.

Tests used in behavioural teratology represent a sensitive instrument to reveal negative effects during its first days of life. When genetic manipulation is the cause of animal suffering, the environmental conditions should be changed or, if this does not relieve the problem, one should seek for alternatives.

References

Altman J, Sudarschan K (1975) Postnatal development of locomotion in the laboratory rat. Animal Behaviour 23, 896-920.

Baxter MR (1983) Ethology in environmental design for animal production. Applied Animal Ethology 9, 207.

Barlow C, Meister B, Lardelli M, Lendahl U, Vennstrom B (1994) Thyroid abnormalities and hepatocellular carcinoma in mice transgenic for v-erbA. EMBO Journal 13 (18), 4241-4250.

Broom DM (1991) Animal welfare: concepts and measurement. Journal of Animal Science 69, 4167-4175.

Bruggemann EP (1993). Environmental safety issues for genetically modified animals. Journal of Animal Science 71(3), 47-50.

Camper SA, Saunders TL, Kendall SK, Keri RA, Seasholtz AF, Gordon DF et al (1995) Implementing transgenic and embryonic stem cell technology to study gene expression, cell-cell interactions and gene function. Biology of Reproduction 52(2), 246-257.

Capecchi MR (1989). Altering the genome by homologus recombination. Science 244, 1288.

Costa P, Turillazzi PG, Bernardo A (1985) Effects of prenatal administration of methyl mercury cloride on behaviour development in mice. In: Heavy metals in the environment (Lekkas TD, ed) pp 691-693.

Cockayne DA, Bodine DM, Cline A, Nienhuis AW, Dunbar CE (1994) Transgenic mice expressing antisense interleukin-3 RNA develop a B-cell lymphoproliferative syndrome or neurologic dysfunction. Blood 84(8), 2699-2710.

FELASA (1994) Recommendation for health monitoring of mouse, rat, guinea pig and rabbit breeding colonies. Laboratory Animals 28, 1-12.

Fox WM (1965) Reflex-ontogeny and behavioural development of the mouse. Animal Behaviour 13, 234-241.

Gerlai R, Marks A, and Roder J (1994) T-maze spontaneous alteration rate is decreased in S100 beta transgenic mice. Behavioral Neuroscience 108(1), 100-106.

Hilakivi-Clarke LA (1994). Overexpression of transforming growth factor alpha in transgenic mice alters non-reproductive, sex related behavioural differences: interaction with gonadal hormones. Behavioral Neuroscience 108(2), 410-417.

Iannaccone PM, Scarpelli DG (1993) Exploring pathogenetic mechanisms using transgenic animals. Annals of Medicine 25(2), 131-138.

Isola LM, Gordon JW (1991) Transgenic animals: A new era in developmental biology and medicine. Biotechnology 16, 3-20.

Melton DW (1994) Gene targeting in the mouse. Bioessays 16(9), 633-638.

Paylor R, Johnson RS, Papaioannou V, Spielgelman BM, Wehner JM (1994) Behavioural assessment of c-fos mutant mice. Brain Research 651(1-2), 275-282.

Pursel VG, Rexroad CE (1993) Status of research with transgenic from animals. Journal of Animal Science 71(3), 10-19.

Rodier P, Reynolds SS, Roberts NW (1979) Behavioural consequences of interference with CNS development in early fetal period. Teratology 19, 327-336.

Ruys T (1991) Handbook of Facilities Planning. Vol. 2: Laboratory Animal Facilities. New York: Van Nostrand Reinhold.

Spyker JM (1975) Behavioural teratology and toxicology. In: Behavioural toxicology plenum (Weiss B, Laties VG, eds). New York, pp 311-344.

Ting CN, Kohrman D, Burgess DL, Boyle A, Altschuler RA, Gholizadeh G, Samuelson LC, Jang W, Meisler MH (1994) Insertional mutation on mouse chromosome 18 with vestibular and craniofacial abnormalities. Genetics 136(1), 247-254.

Use of Transgenic Animals and Welfare Implications

M. van der Meer and L. F. M. van Zutphen

Summary

This paper presents the results of a literature survey and an inquiry intended to gain insight into the extent of use of transgenic animals in EU-Member States and into the potential effects of transgenesis on the health and welfare of the animals. Screening of the Medline database revealed data for quantifying research with transgenic animals and for estimating the number of institutes working with these animal models. Questionnaires have been sent to the national contact points (NCPs, i.e. governmental representatives for legislative issues on animal protection) of the EU-Member States. From the responses to the questionnaires and the information available in the literature, the conclusion could be reached that the degree of suffering caused by the experimental procedures and insertional mutations is, in general, less severe than the welfare problems caused by the direct effect of the expression of the transgene. However, only very few data are presently available to substantiate this conclusion. Further research is necessary to identify and quantify indicators of discomfort in the animals involved in the process of transgenesis. A proposal is made for the development of a model protocol, to be used as a standard operating procedure for evaluating adverse effects in the process of producing transgenic animals.

Introduction

The early work on transgenic animals mainly involved mice and these animals continue to be the most commonly-used species. Since 1980, thousands of transgenic mice have been produced and used for biomedical research (Lathe and Mullins 1993, Buehr and Hjorth 1994, Monastersky and Robl 1995). The novel methods of genetic modification also provided the opportunity to introduce specific changes into the genome of livestock, and although the efficiency is low, successful gene transfer has been reported for pigs, sheep, goat and dairy cattle (Pursel and Rexroad 1993, Ebert and Schindler 1993).

The production of transgenic animals has raised some controversial issues. On the one hand, transgenic animals are more refined animal models than those in which genetic changes are achieved by traditional breeding methods. Genetic engineering is not a simple extension of previously existing techniques. It allows not only the insertion into the genome of new or extra copies of genes, which may be derived from a different species, but also the "knock out" of targeted genes (Hubrecht 1995). These methods provide very precise tools to investigate the function of a single gene *in vivo*. By transgenesis it is also possible to produce animal models of human diseases for which no natural models are available. The increased efficiency of research with transgenic animals could lead to a diminished use of some experimental animals. There is also the potential for the commercial use of the results of transgenic research, such as the production of essential human proteins in the milk of transgenic livestock.

On the other hand, there is also substantial public concern about these animals. Questions have been raised dealing with the safety aspects of this new development. In particular, fear has been expressed of ecological disasters due to the escape of transgenic animals from laboratories into the wild. Because of the fact that genetic material can be exchanged between different species, it also has become a major ethical issue. How far can we go and who is imposing the limitations? Is everything which is possible technically also acceptable ethically? And what are the welfare problems of transgenic animals? Welfare problems may arise that go beyond those caused by the effect of the genetic trait which is introduced or changed. The two techniques used most frequently for inserting genetic material into the mammalian germ line are pronuclear microinjection and embryonic stem (ES) cell-mediated gene transfer. The random integration of the DNA into the genome by microinjection may increase the chance of disturbing normal physiological processes of the animal. Therefore one could reason that there is an increased chance of welfare problems in transgenic animals when compared to an animal model, in which traits are introduced by traditional breeding methods. However, specific studies aiming at quantifying the degree of discomfort caused by the technique of transgenesis are still scarce.

In order to obtain more insight into some of these aspects, data have been collected during 1995 as part of a project funded by the European Commission. The project has been carried out within the scope of the research programme on Socio-Economic Impact Studies of Biotechnology of the European Commission (BIOTECH programme DGXII) and aims to provide insight into the extent of use of transgenic animals in the EU-Member States and to reveal information on welfare studies in this area of research. The project should also generate information for the design of a study for the development of a model protocol for evaluating adverse effects during the process of transgenesis.

The methodological approach of the project included a literature survey and inquiring of the national contact points (NCPs; the governmental representatives for legislative issues on animal protection) of the different EU-countries.

An overview of the main results is presented in this paper, while the detailed results of this EC-project are described in a separate report (Van der Meer 1996).

Transgenic Research in the EU-Member States

The questionnaire for the NCPs included the following items:
1) the number of institutes involved in the production and/or use of transgenic animals;
2) the aims of research;
3) the number of animals, per species, used for transgenic research;
4) animal welfare problems noticed by researchers, and
5) welfare studies in this area of research.

Although most of the NCPs responded to the questionnaire, only limited information was obtained by this approach. The main problem is that in most countries there is no separate registration of transgenic experiments. Therefore, most of the quantitative data were collected by performing a literature survey. Medline, a literature database produced by the US National Library of Medicine USA and covering over 3,400 international biomedical journals, was used. This database also includes the addresses of the authors' institutes. This made it possible to obtain an overview of the institutes performing transgenic research in each of the EU-Member States.

Although the database does not cover all research performed with transgenic animals (as not all results will be published in international scientific journals), the analysis of the results of this searching will at least provide a reliable indication as to the ratio of transgenic research between the different EU-countries.

Figure 1 gives an overview of the number of papers on transgenic animals published by institutes from EU-countries (+ Switzerland) (Medline search period 1990-1995). The number of published papers also includes review papers.

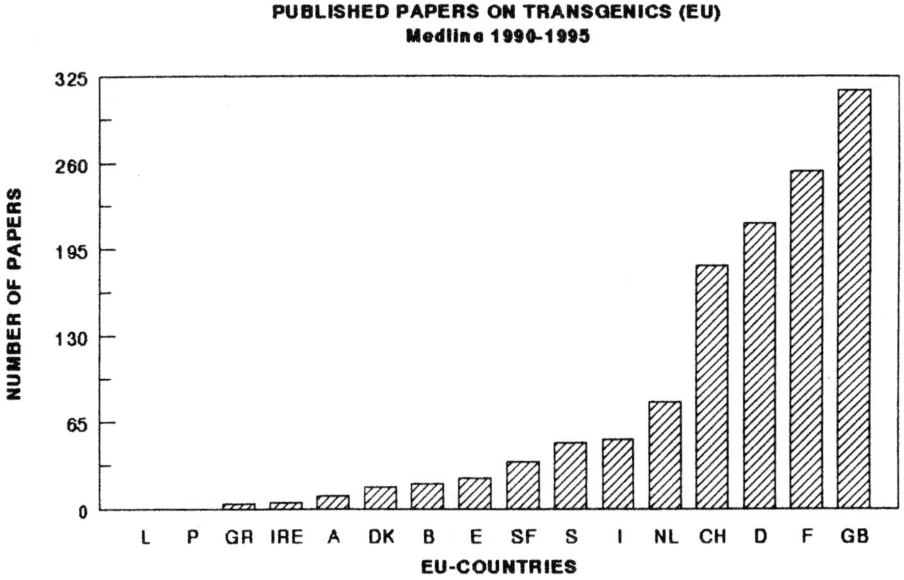

Figure 1: *Total number of papers on transgenic animals as published by institutes from different EU-countries (+ Switzerland) over the period 1990-1995.*
(L=Luxembourg, P=Portugal, GR=Greece. IRE=Ireland, A=Austria, DK=Denmark, B=Belgium, E=Spain, SF=Finland, S=Sweden, I=Italy, NL=Netherlands, CH=Switzerland,

The total number of papers on transgenic animals as published by scientists from EU-Member States, found in the Medline search 1990-1995, was 1075. The data indicate that of the EU-countries, Great Britain, France and Germany are the most productive in the field of transgenic research.

The number of institutes involved in transgenic research in the EU-countries can also be estimated from the search in Medline 1990-1995. A limitation of this approach is that institutes, reporting in the beginning of this search period may already have stopped their research and also other institutes may have started, but not published yet. Therefore up to date information provided by NCPs has also been included when available. Each university or private company is counted as one institute, although there may be several departments performing transgenic research.

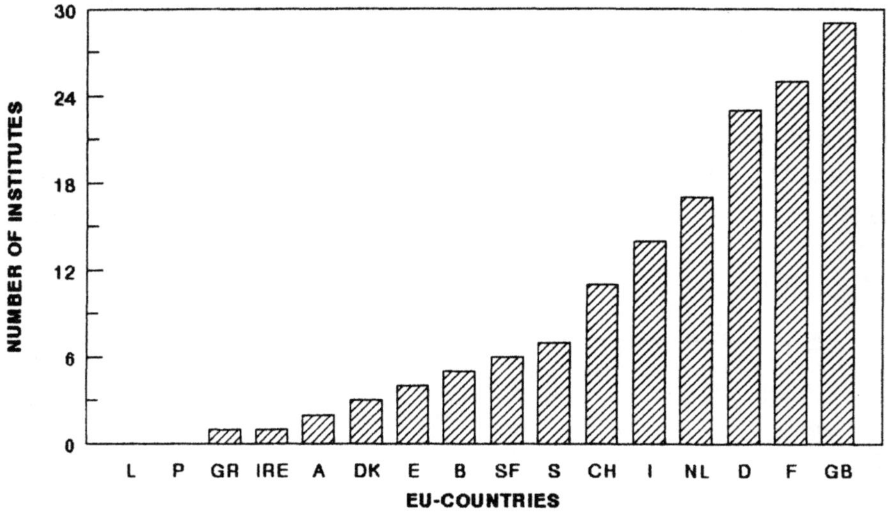

Figure 2: *Total number of institutes producting and/or using transgenic animals in EU-countries (+ Switzerland) during the period 1990-1995.*

Only very few data are available as to the number of animals used in transgenic research.

Table 1 illustrates the number of animals (per species) used for transgenic research, according to the information obtained from the NCPs of five countries. The data shown for Finland and Sweden were supplied with the remark that these are rather rough estimates. The data of Switzerland relate to the number of mice used for the production of transgenic animals, the offspring of the existing transgenic strains are not included.

Table 1. Number of animals used for transgenic research in five countries (see text)

ANIMAL SPECIES	Netherlands (1991)	Finland (1994)	Switzerland (1994)	Great Britain (1994)	Sweden (1994)
MOUSE	62,665	ca.2,500	12,919	181,204	ca.32,665
RAT	25	ca.500	-	2,313	ca.3,265
LIVESTOCK	39	ca.50	-	619	ca.17
BIRD	-	-	-	52	ca.330

From the data of Table 1, it is obvious that the mouse is still the species used most frequently for transgenic research. The main fields of research are the study of gene expression, immunology, embryology, tumour development, neurology, human disease models and testing genes for the production of proteins in the milk of livestock. Rats are used mainly as models for hypertension studies. Because rats are important animals in the study of the cardiovascular system, several lines of transgenic rats have been produced expressing genes involved in the regulation of blood pressure. Transgenic rat models are being developed for testing the efficiency of drugs and other therapies on reducing high blood pressure. Transgenic livestock have mainly been produced as bioreactors, for instance cattle are being used for the production of human lactoferrin, collagen or erythropoietin. Transgenic sheep are producing alpha-1-antitrypsin in their milk. Transgenic pigs have been produced with human growth hormone and are also being developed as potential organ donors for xenotransplantation studies. Transgenic birds (chickens) are used, for instance for disease resistance research, whereas transgenic fish have been used for the study of basic gene regulation and expression.

Welfare Aspects of Transgenic Animals

The process of transgenesis can cause welfare problems, beyond those that are caused by the effect of the genetic trait which is introduced or changed. Most transgenic animals are still being produced by the microinjection method. This method of transgenesis can cause effects on the health and welfare of animals at different levels (Van Reenen and Blokhuis 1993):

- *At the level of the experimental procedures that are necessary to introduce the DNA construct into the zygote.* The various methods and techniques used for the introduction of the DNA construct into the zygote may have implications for animal welfare. Microinjection can cause mechanical damage to the DNA of the embryo. The production of transgenic animals also involves embryo culture *in vitro* with the transfer of the altered embryos into female animals to complete their development. Each of these treatments may affect the normal development, and thus the health and welfare, of the individual. The parental animals (donor animals, vasectomized males, foster mothers) which are used for the production of the transgenic progeny can also experience discomfort as a result of the experimental procedures, such as anaesthesia and injections.
- *At the level of integration of the microinjected DNA into the genome.* After microinjection, the transgene is integrated randomly into the host genome, meaning that the transgene may be inserted anywhere within the

host chromosomes. Foreign DNA may disrupt the expression of a natural gene and an insertional mutation can occur. If such mutations occur in physiologically important endogenous genes, there could be subsequent adverse effects on animal welfare.
- *At the level of expression of the introduced gene.* Welfare problems can also occur either as a direct result or as a side effect of the expression of the transgene. The severity of the detrimental effects depends on the nature of the deviation caused by the transgene.

From the literature survey, it appears that the extent of research specifically investigating the effects of transgenesis on the welfare of animals is rather limited. Only a few papers on this topic have been published so far (Van Reenen and Blokhuis 1993, Broom 1993, Moore and Mepham 1995, Poole 1995, Hubrecht 1995). The information obtained from the NCPs confirmed that research aiming at welfare aspects of transgenic animals is scarce.

Based on the data collected for this study, an estimate can be made as to the discomfort experienced by mice (parental animals) as a result of the experimental procedures involved in the process of transgenesis. Mice are often housed individually during the time of the procedure (especially the vasectomized males). Solitary housing of mice is known to cause discomfort. Most of the female mice used for superovulation are rather young when first mated (at the age of 3-4 weeks old) and the procedure may involve some stress (Hubrecht 1995). These donor mice also receive two hormonal treatments with gonadotrophins, which increases the number of eggs that the ovary releases. The superovulated females are mated with fertile males. The fertilized eggs are collected from these donor mice. These animals are killed humanely (a quick and painless death).

Male animals are submitted to a surgical procedure under general anaesthesia to be vasectomized (transection of the vas deferens). Recovery from general anaesthesia is estimated as moderate discomfort. They are then mated with the foster mothers to induce pseudopregnancy.

The foster mothers are also submitted to a surgical procedure under general anaesthesia for the implantation of the microinjected oocytes. With a fine glass needle the embryos are inserted via the infundibulum into the oviduct of the foster mothers. After nursing the progeny, the foster mothers are killed humanely. The non-transgenic offspring and the chimaeras are also often euthanized.

At present, no reliable information is available as to the potential adverse effects of the manipulations with the oocyte or the embryo for the progeny. Gene transfer by microinjection is not a procedure with a predictable outcome, but rather a trial and error procedure with many unsuccessful attempts and outcomes. It can diminish the viability of the embryos, for instance by mechanical damaging the DNA of the zygote, or because there is usually no control over where the gene inserts itself into the genome of the recipient, or over the number of copies of the gene which become incorporated.

When a transgene integrates into the genome, the construct may disturb the expression of the natural genes and an insertional mutation can occur. According to the literature, these unintended mutations occur in 5-15% of the total number of transgenics after microinjection (Constantini *et al.* 1989, Meisler 1992). These frequencies may be higher, because some mutations can be lethal at an early stage, while others stay undetected (internal or behavioural) or manifest themselves only after continued breeding (homozygotes). If such mutations occur in physiologically important endogenous genes, there could be a subsequent adverse effect on the animal's welfare. The degree of discomfort will depend on the function of the mutated gene.

Besides microinjection, the production of transgenic animals also involves culturing embryos *in vitro* and the reimplantation of these embryos into the foster mothers. Each of these treatments may affect the normal development and thus the health and welfare of the animal. However, the discomfort for the progeny caused by these treatments is difficult to establish, because the animals which are less vital may not be born or the weak animals may be killed soon after birth by their mother (a normal process for weak mice). Because such maternal killing appears to occur more often with transgenic animals than with non-transgenic animals, this might indicate that the viability is impaired as a result of the transgenic technique.

Most evident welfare problems seem to occur as a direct result or as a side effect of the expression of the transgene. The extent to which the expression of the introduced gene construct may lead to detrimental effects depends on the nature of the deviation caused by the transgene. Substantial discomfort may be expected to occur in the production of transgenic animals for some human disease models or cancer research. This is not essentially different from the discomfort caused in animal models produced according to traditional breeding techniques. However, examples have been described where the expression of the gene has severe side effects, so that the welfare is seriously compromised. This can be illustrated by the so called "giant mouse", which has been genetically engineered using a human growth hormone gene (hGH). These mice are not only larger and heavier than normal laboratory mice. They also show severe side effects, like chronic kidney and liver disfunction, tumour development, structural changes in heart and spleen, high infant and juvenile mortality, shortened lifespan and reduced fertility (Brem and Wanke 1988, Berlanga *et al.* 1993, see also chapter of Wolf and Wanke in these Proceedings). In a recent paper Poole (1995) discussed the ethical and scientific implications of the use of these animals. He indicated that the quality of science and the value of the likely outcome should be balanced against the suffering inflicted on the animal. Since the welfare of the hGH mouse is seriously compromised Poole questioned whether this transgenic animal can be considered as an appropriate model. Wolf and Wanke have indicated that the pathogenesis of glomerulosclerosis and the mechanisms involved in hepatocarcinogenesis can adequately be studied in hGH transgenic mice. Also, therapeutic strategies can be tested before the animals suffer from the consequences of renal insufficiency. They stated that the well-being of these hGH transgenic mice may be less disturbed than that of respective conventional animals used for studying these problems. The multiple pathologies described above occur when the animal ages, but when using young transgenic animals for this research, the suffering of the animals might be limited.

Another example of a transgenic animal in which the normal development is seriously disrupted is the "legless mouse". A construct, consisting of a fruitfly gene (heat shock protein) and a viral gene (thymidine kinase) has been inserted into mice (McNeish *et al.* 1988, 1990). The heterozygous mice were normal, but after inbreeding homozygous mice were born with major developmental abnormalities, including the loss of hindlimbs, malformed front limbs, facial clefts and massive brain defects. These animals died within 24 hours after birth. The transgene had disturbed the part of the genome, which plays an important role in the development of limbs and brains. Although the malformations arose as side effects, they now became subjected to embryogenetic studies.

Several other examples with various degrees of discomfort, due to direct or indirect effects of the expression of the transgene are described in the literature (Appleby 1997, Archibald *et al.* 1990, Janne *et al.* 1994).

Conclusions and Recommendations

The use of transgenic animals for research purposes has increased substantially during the last decades. Many new models have been developed. Transgenic animals seem to be indispensable for progress of the research in several biomedical disciplines.

The process of transgenesis may cause welfare problems for the animals. However, based on the present insights, it is usually not possible to give a well-founded judgement on the degree of discomfort caused by transgenesis.

In order to gain more insight into the potential welfare problems caused by transgenesis, further research is necessary in particular to identify and quantify indicators of discomfort. Therefore, we have developed a model protocol that could possibly be used as a standard for systematic observations of animals and for collecting information which might be relevant for the (retrospective) estimate of the impact of transgenesis on the well-being of animals. The feasibility of this model protocol is the subject of study in a controlled project.

Short Outline of the Feasibility Study
Mice, produced in current transgenic experiments, will be observed and screened for a number of physiological/clinical and ethological parameters which are frequently used in studies aiming at quantifying the degree of discomfort. Also *post mortem* examinations will be performed. The effects of genetic modification will be studied in different transgenic experiments. Within each of these experiments four groups of animals, all derived from the same inbred strain, will be compared:

1) control animals,
2) transgenic animals,
3) non-transgenic animals after transgenesis, and
4) animals with a nonsense (non-functional) transgene.

The objective is to select the potential indicators of discomfort that can be monitored on a routine basis (for instance by animal husbandry staff and technicians) in the process of producing transgenic animals. Parameters to be included into the model protocol should be sensitive, non-invasive and easy to determine. In this project several parameters will be tested. In order to evaluate the value of a given parameter for inclusion into the model protocol, the time needed to collect the data will also be established for each parameter. Parameters to be tested include for instance survival rate, weight curve, developmental parameters (incisor eruption; eyes open; ears open; hair growth), clinical parameters (body posture; coat condition; eye-/nose excretion) and behavioural parameters (grooming, climbing, eating, drinking, locomotion, resting). After death, the animals will be examined by an experienced pathologist.

All data from each of the different groups will be collected and analyzed statistically in order to quantify whether there are any significant differences. The differences which will be found between the four groups will serve as the basis for the estimation of the sensitivity of the parameters and the impact of transgenesis on the welfare of animals. Subsequently a selection of parameters will be made to establish the final model protocol which should provide a practical tool for quantifying indicators of discomfort. This selection will be based on criteria of feasibility and sensitivity. The parameters used for the final model protocol should be practicable and non-invasive. The final model protocol should include the time scales and equipment necessary to perform the different measurements. The project starts in 1996 and will be performed at the Department of Laboratory Animal Science, Utrecht University.

References

Appleby MC (1997) Genetic engineering and welfare. Journal of Applied Animal Welfare Science, in press.

Archibald AL, McClenaghan M, Hornsey V, Simons JP, Clark AJ (1990) High-level expression of biologically active human α-1-antitrypsin in the milk of transgenic mice. Proceedings of the National Academy of Sciences of the USA 87, 5178-5182.

Berlanga J, Infante J, Capo V, Delafuente J, Castro FO (1993) Characterisation of transgenic mice lineages. 1. Overexpression of hGH causes the formation of liver intranuclear pseudoinclusion bodies and renal and hepatic injury. Acta Biotechnologica 13, 361-371.

Brem G, Wanke R (1988) Phenotypic and patho-morphological characteristics in a halfsib-family of transgenic mice carrying foreign MT-hGH genes. In: New Developments in Biosciences: Their Implications for Laboratory Animal Science (Beynen AC, Solleveld HA, eds). Dordrecht: Marinus Nijhoff, pp 93-98.

Broom DM (1993) Assessing the welfare of modified or treated animals. Livestock Production Science 36, 39-54.

Buehr M, Hjorth JP (1994) Genetically modified animals: perspectives in development and use. Miljoproject nr. 277, Ministry of the Environment and Energy, Danish Environmental Protection Agency, Copenhagen K.

Constantini F, Radice G, Lee JJ, Chada KK, Perry W, Jin Son H (1989) Insertional mutations in transgenic mice. Progress in Nucleic Acid Research and Molecular Biology 36, 159-169.

Ebert KM, Schindler JES (1993) Transgenic farm animals: Progress report. Theriogenology 39, 121-135.

Hubrecht R (1995) Genetically modified animals, welfare and UK legislation. Animal Welfare 4, 163-170.

Janne J, Hyttinen JM, Peura T, Tolvanen M, Alhonen L, Sinervirta R, Halmekyto M (1994) Mini-review: Transgenic bioreactors. International Journal of Biochemistry 26, 859-870.

Lathe R, Mullins JJ (1993) Transgenic animals as models for human disease. Report of an EC Study Group. Transgenic Research 2, 286-299.

McNeish JD, Scott WJ, Potter SS (1988) Legless, a novel mutation found in PHT-1 transgenic mice. Science 241, 837-839.

McNeish JD, Thayer J, Walling K, Sulik KK, Potter SS, Scott WJ (1990) Phenotypic characterization of the transgenic mouse insertional mutation: Legless. Journal of Experimental Zoology 253, 151-162.

Meisler MH (1992) Insertional mutation of "classical" and novel genes in transgenic mice. Trends in Genetics 8(10), 341-344.

Monastersky GM, Robl JM (1995) Strategies in transgenic animal science (xviii). Washington DC: ASM Press.

Moore CJ, Mepham TB (1995) Transgenesis and animal welfare. Alternatives To Laboratory Animals (ATLA) 23, 380-397.

Poole TB (1995) Welfare considerations with regard to transgenic animals. Animal Welfare 4, 81-85.

Pursel VG, Rexroad CE (1993) Status of research with transgenic farm animals. Journal of Animal Science 71 Supplement 3, 10-19.

Van Reenen CG, Blokhuis HJ (1993) Investigating welfare of dairy calves involved in genetic modification: problems and perspectives. Livestock Production Science 36, 81-90.

Van der Meer M (1996) Transgenic animals and animal welfare. EC-Report, Department of Laboratory Animal Science, Utrecht University.

Transgenic Animals: The Need for Ethical Dialogue

P. Sandøe, B. Forsman and A. K. Hansen

Summary

To meet the sceptical attitudes of the public, those involved in the development and use of transgenic animals may take the approach of providing as much factual information as possible about animal biotechnology, called here the information approach.

The main aim of this paper is to argue that the information approach cannot stand alone, and to present a broader approach, the approach of ethical dialogue, which explicitly takes value questions into account.

In presenting this approach, a number of common ethical concerns about the development and use of transgenic animals are formulated. One of these, the concern for the genetic integrity of the animals, is discussed in more detail. An alternative ethical view, based on the concern for animal welfare, is outlined and is applied to the various uses of transgenic animals.

Introduction

The general public is in various degrees sceptical about the development and use of animal biotechnology (Tait 1990).

Those involved in research and development relating to transgenic animals will have, in most cases, a positive attitude to their own work. They may therefore be tempted to think that the only thing which prevents the general public from sharing their own positive attitude is lack of information. Consequently, to meet the sceptical attitudes of the public they may take the approach of providing as much factual information as possible about animal biotechnology. We shall call this the information approach.

Taken literally, the information approach says that one should give the general public as much information as possible about transgenic animals.

However, although access to all relevant information is necessary for participation in a meaningful discussion, much of the information available concerning transgenic animals will simply not be of any interest to the general public. To get information to the public, for example via the media, one has to be selective. If the aim is to alleviate the worries of the public, the information given must address these worries, and therefore, in practice, the information approach is not always feasible.

Our own experience from being involved in public discussions concerning animal biotechnology is that there is often a marked difference between what a scientist involved in research or development of animal biotechnology will perceive as ethically relevant and what worries the interested lay person. For example, the scientist may provide information concerning possible costs and benefits, potential risks to the environment and to human health, and consequences for animal welfare, but as these may not be the main worries of the general public, the information may not serve its purpose. Quite to the contrary, the scientist may come across as someone who is simply not sensitive to what matters from an ethical point of view. If the scientist wants to influence the public perception of transgenic animals, he must therefore start by trying to understand the values which underlie the worries of the public, and not just address those worries that he himself regards as legitimate. This calls for a different approach.

The Approach of Ethical Dialogue

This approach involves the following four steps:

1. Formulating the concerns to which transgenic animals give rise.
2. A critical discussion of these concerns.
3. To state a set of ethical considerations which can stand up to scrutiny.
4. To apply the stated ethical considerations to issues regarding the development and use of transgenic animals.

1. Formulating the concerns to which transgenic animals give rise

The first step in establishing an ethical dialogue must be to formulate the different concerns to which the development and use of transgenic animals give rise and the underlying ethical values or principles. Three such concerns have already been touched upon, i.e.

(a) that too many resources are spent on transgenic animals;
(b) that there may be risks to the environment and to human health; and
(c) animal welfare.

Nearly everyone would probably agree that the values underlying these concerns are ethically significant, but these may not be the only concerns. Thus according to Donnelley *et al.* (1994):

> "We usually consider only individual animal welfare or experimental well-being, particularly pain and suffering, and not the broader significance of animal and animate life *per se*. But the new animal biotechnologies do not easily let us off the hook. Their practical implications go decidedly beyond the more familiar ethical issues of individual welfare and well-being. Certain new biotechniques delve deeply into and manipulate the very character of animal and animate being. These new technologies have the potential to alter nature radically."

Such other concerns have been voiced in the public debate concerning transgenic farm animals, for instance that

(d) the genetic integrity of animals will not be respected;
(e) the technology may lead to animals being treated as mere things, rather than as individual subjects of conscience;
(g) "unnatural" animals may be created; and
(h) the techniques may be misused on human beings.

These will not always be presented as separate concerns, but may appear mixed with one another.

It is most important to be aware that even though most people seem to share some of these ethical values, especially the concerns of a, b, and c, they may attach quite different weight to them. For example, two persons may agree that both benefits to humans and animal welfare matter, but they may differ radically when it comes to balancing economy against animal welfare. One person may claim that we may compromise animal welfare only if it is a matter of life and death for humans, whereas another may think that lower consumer prices can justify considerable limitations in the welfare of farm animals.

2. A critical discussion of the concerns

The second step of the ethical dialogue should consist of a critical discussion of the formulated concerns and the underlying ethical values or principles. This critical discussion should spell out clearly the principle involved for each ethical consideration, and it should try to establish the consequences of taking this principle seriously in all its instances.

Take as an example the ordinary reaction against genetic engineering of farm animals[1]. Many people resent the idea of changing the genes of animals by means of genetic engineering. For example, at a so-called consensus conference held in Copenhagen in September 1992 a panel of lay persons concluded that animal welfare is not the only criterion for deciding the ethical acceptability of genetic changes in animals. They also thought it unethical to induce genetic changes in domestic animals to adjust them to existing agricultural production methods. Thus, the members of the panel seemed to view it as an ethical duty not to change the genetic make-up of domestic animals. It should be noted that the panel did not reject all instances of genetic engineering, for instance in the view of the panel, transgenesis as part of vital biomedical research might be acceptable. However, the panel rejected the use of genetic engineering merely to give us cheaper food products. One possible interpretation of this conclusion is that the panel ascribes a negative value to genetic engineering of animals - a value which may only be counterbalanced when vital needs are at stake.

The ethical concern underlying this resentment may be formulated as a demand to respect the genetic make-up of animal species. Vorstenbosch (1993) defines the concept of genetic integrity in the following manner:

> "In biotechnology the genetic integrity of the individual animal and of the species is central. We can define the genetic integrity of the animal as the genome being left intact. This seems to be a meaningful notion in view of the fact that we clearly point out some factors or actions by which the genome would not be left intact."

A similar concept also seems to be in the mind of Michael Fox (1990) when he reacts against the possibility of developing transgenic animals:

> "Genetic engineering makes it possible to breach the genetic boundaries that normally separate the genetic material of totally unrelated species. This means that the *telos*, or inherent nature, of animals can be so drastically modified (for example by inserting elephant growth hormone genes into cattle) as to

1 The following is not meant to be an exhaustive ethical analysis, but should rather serve to illustrate how a critical discussion of ethical values may proceed; for a more detailed account see Sandøe and Holtug (1993) and Sandøe, Holtug and Simonsen (1996).

> radically change the entire direction of evolution, and primarily toward human ends at that. Is that aspect of the animal's *telos* we refer to as the genome and the gene pool of each species not to be respected and not worthy of moral consideration?"

Fox does not speak about genetic integrity but about respecting the *telos* of an animal species, though the two things seem to come to the same thing[2]. His view represents a type of ethics, which is concerned about the fate of whole species of animals rather than about the lot of individual animals. In contrast, the type of ethics represented by concerns about animal welfare only focuses on the individual animal.

There is no doubt that the idea of genetic integrity captures a worry that a great many people have. Still, there is scope for a critical discussion of the ethical significance of genetic integrity. The demand to respect the genetic integrity of animals is based upon a reaction against genetic engineering. However, as a logical consequence, this demand will also apply to selective breeding, as like transgenesis, selective breeding will result in significant genetic changes in animal species. If the argument against transgenesis is that genes are being changed significantly, it will just as well be an argument against selective breeding.

It may be asked what is so special about the genetic structures which exist right now. Through the history of evolution the genetic structures have changed continuously. It is not possible to point to one stage in evolution at which animal species have reached their "final" or "real" development. To say that the present genetic make-up is special seems completely arbitrary from the perspective of evolution - just like saying that art and literature have now reached their final points and should not be allowed to change further.

---

2 It should be noted that the notion of an animal's telos does not by itself imply that it is wrong to change that telos. The notion has been framed by Bernard Rollin (1986) in an attempt to clarify what animal welfare is. To violate the *telos* of an animal, according to Rollin, one must prevent the animal from fulfilling its natural desires or inclinations. In fact, Rollin suggests that it is not by itself wrong to alter the *telos* of an animal; this is only wrong if it implies an infringement of the new *telos*. Thus, according to Rollin "... it is only wrong to change a *telos* if the individual animals of that sort are likely to be more unhappy or suffer more after the change than before." Rollin's view appears to coincide with the animal welfare view defended below. Fox, on the other hand, thinks that the genetic integrity of animal species ought to be protected and he therefore thinks that we always do something bad when we change the *telos* of an animal species.

Another question that could be asked about the idea of genetic integrity is whether genetic interventions for certain purposes should not be viewed as ethically admirable. Thus breeding for increased health may be viewed simply as a good thing. It may be breeding of hens resistant to Marek's disease or the breeding of pigs without malignant hyperthermia. Although this kind of breeding may be seen as a remedy of damages in the gene-pool established during genetic selection of high-yielding domestic species, such breeding may effectively eradicate genotypes disposed to certain illnesses and thereby secure the health of domestic stock. If the idea of genetic integrity is valid, however, it is not a good thing to breed for increased health or at most it is the lesser of two evils. This ought to worry the proponent of genetic integrity.

Finally, it could be asked how respect for genetic integrity benefits the animals. Respecting, or not respecting, genetic integrity will both have consequences for the kind of animals we bring into existence. For example, if we respect genetic integrity we will be more reluctant to start programs of selective breeding and more disposed to maintain existing breeds. Does this always benefit the animals? We think that the answer to this question must be negative. The only difference that respect for integrity will make, concerns which animals benefit and which are being harmed, and not whether animals benefit or are harmed. The net result in terms of harms and benefits may be the same or even negative.

It might be replied that often there is a difference in that some strains of animals experience a higher level of welfare than others, and that therefore it may be more beneficial to develop one kind of animal rather than another. However, this answer seems to point in quite a different direction: There is no guarantee that the genetic structures which exist right now are more conductive to animal welfare than others which may be bred or be developed by means of genetic engineering. Rather, this point speaks in favour of saying that the genuine worry underlying talk about genetic integrity is really for animal welfare.

The critical discussion of genetic integrity should, of course, not end here. Thus the adherent of genetic integrity should have a chance to answer the critical questions raised. For example, the proponent of animal integrity may say that what, on reflection, worries him concerning transgenic animals is not the mere fact that genes are being changed but that species barriers can be broken or that the velocity of changes occurs at a different order of magnitude. The answers may prompt new questions and so on. In the end there may be three outcomes: First, that one party breaks off the dialogue - with high potential costs to his credibility. Secondly, that a consensus is reached. Thirdly, that the parties agree to disagree, though understanding why the other party has a different opinion.

As the example hopefully shows, there may sometimes be plenty of scope for arguing about the validity of ethical considerations. The arguments should give rise to an open dialogue where the starting point is an attempt to take seriously the other person's ethical considerations. This, of course, implies that the person who engages in the ethical dialogue cannot himself avoid getting involved in ethical thinking.

3. Stating a set of ethical considerations which can stand up to scrutiny

The third step of the ethical dialogue must, in the light of the previous critical discussion, be to formulate a set of ethical considerations which can stand up to scrutiny. For example, human health, environmental risks and animal welfare may be aspects which should be considered when deciding under which circumstances it is ethically acceptable to develop and use transgenic animals.

Starting with animal welfare, transgenic animals may be assessed through the following two steps: (1) finding out what the realistic alternatives are (including leaving things as they are); and (2) assessing the likely positive and negative effects on animal welfare (without forgetting the possibility of unforeseen negative effects).

These two steps, of course, cannot stand alone. To reach an ethical conclusion two further steps must be taken: (3) finding out which benefits are at stake; and in case animal welfare is adversely affected, 4) finding out whether the "costs" in animal welfare will outweigh the human benefits.

This "welfare approach" has two obvious advantages over genetic integrity. First, what matters ethically according to the welfare approach is by definition what matters to the animals. Therefore, there is no problem explaining how this approach benefits the animals. According to the welfare approach one should, other things being equal, aim only to bring into existence animals which have a good life.

The other advantage is that ethical assessments of transgenic animals will be made in the same way as with other "animal welfare issues", for instance intensive livestock production and animal research. To decide whether there is an ethical problem one should look for the same things in all cases: abnormal behaviour, physiological stress-reactions, health problems, etc. In that sense, there is nothing special about the development and use of transgenic animals.

This is not to say that it is a simple matter to draw ethical conclusions on the basis of the welfare approach. On the contrary, reaching an ethical conclusion involves a whole series of steps, not all of which are simple and straightforward. For example, it must be decided what matters to an animal; is it only avoiding suffering, or do pleasure and other "positive" experiences also matter? Also, among other things, it must be decided how the scientific facts concerning behaviour, physiology and health bear on what matters to the animal (Sandøe and Simonsen 1992).

An important point to be made is that despite many complications and controversial issues the welfare approach is already at work, for instance as a basis for the regulations and recommendations issued by the EU and the Council of Europe, as well as national legislation.

Not everyone may agree with the conclusions drawn at this step. However, even if people do not agree, they may recognize that the ethical views presented are based upon a process of reflection where alternative points of view are considered in a serious way.

4. Applying the ethical considerations

Finally, at the fourth step the ethical considerations should be applied to the specific questions about the development and use of transgenic animals which initally gave rise to the ethical discussion. We suggest here a policy for the assessment of transgenic animals which is in line with the "welfare approach" outlined above. This policy is based on the following principle: To justify the development and use of transgenic animals, vital interests should be at stake.

Attempts to change animals by means of genetic engineering often have unforeseen negative consequences on animal welfare. An extreme example is provided by transgenic pigs producing human or bovine growth hormone. These animals suffered from a number of serious health problems, including lameness, susceptibility to stress, peptic ulcers, parakeratosis, lethargy, anoestrus in gilts and lack of libido in boars (Pursel *et al.* 1989).

Some of these unforeseen consequences could probably have been foreseen if a careful risk assessment had been undertaken. For example, it could have been predicted that animals with continuously elevated levels of growth hormone will experience health problems, especially leg disorders. Therefore, assessments of risks concerning animal welfare should precede all attempts to change animals either through selective breeding or by means of genetic engineering.

However, far from all the consequences for animal welfare can be foreseen as negative. Therefore the expected gains (for humans and animals) should be considerable to justify attempts to make radical genetic changes in animals.

This tells against using transgenic techniques merely to improve the productivity of farm animals. There is no reason to believe, that a general increase in productivity of farm animals in the Western world will benefit the farmers or feed the starving people of the world. It will in the long run only lead to lower prices, although some farmers will benefit in the short run. Lower prices on animal products is not an urgent need in our part of the world. To substantiate this bold claim it may be mentioned that the average person in Denmark spends less than 13 % of the income after tax on food - which is less than half of the amount spent 30 years ago.

Therefore, there are good ethical reasons in favour of international regulations against animals "designed" to increase productivity. If such regulations are in force for all those who compete in the same market they will not affect the farmers, and since the consumers spend relatively little on food they will not be affected in a significant way.

A case where vital human interests are at stake may be provided by the use of transgenic livestock for the production of therapeutic protein. For example, the company PPL Therapeutics now has a flock of transgenic sheep which produce AAT (alpha-1-antitrypsin) in their milk. The protein is expressed in large quantities, about 15 g per litre milk (Buehr and Hjorth 1994). AAT is a vital drug for the treatment of patients suffering from emphysema, a lung condition which is potentially life-threatening. This pharmaceutical is at present derived from human plasma. However, it is difficult to get AAT in sufficient quantities, and therefore the treatment is not available to all who need it. Also AAT produced from the transgenic sheep will be much cheaper and more safe. The transgenic sheep do not seem to have any welfare problems due to their genetic modifications (but, of course, the animals used in the experimental phase and during the process of producing the transgenic animals may have experienced some suffering). Since the animals will satisfy a vital human need and since there seems to be no major welfare problems for the animals, a positive attitude seems to be required from the point of view of the welfare approach. A similar line of argument seems to apply to the development of transgenic animals to be used as organ donors.

A more pressing, but also more difficult problem is raised by the development and use of transgenic laboratory animals. On the one hand such animals may serve vital human purposes, holding out promising prospects for future developments in human medicine when used as models of human diseases, for the testing of products to which humans will be exposed and in basic research. On the other hand, these animals may experience reduced welfare ranging from mild discomfort to intense suffering.

This is, of course, an instance of a general dilemma concerning those animals used in experiments which are on the one hand vital to humans and on the other hand harmful to the animals. Presently, we are referring only to experiments which we all have reason to believe may give us some new and useful results, which it is not possible to perform without causing serious harm to some of the animals involved, and where the same results could not be achieved by alternative methods, or by causing less harm to the individual animal or by experiments harming a smaller number of animals[3].

From the point of view of the welfare approach, the ethical issue is not different just because the experiments involve the development or use of transgenic animals. Of course, transgenic disease models may require special care and monitoring, which is not required for normal healthy animals. However, this is also the same for disease models developed by breeding spontaneously-occurring mutants or by some chemical or surgical induction procedures. So, again, transgenic animals do not raise an ethical problem that is in principle special or new. Of course, the transgenic animal may appear special because it opens our eyes to the relevant problems for the first time.

The policy for an ethical assessment of transgenic animals here outlined may not be accepted by all those engaged in the public debate concerning transgenic animals. But, at least, it may be recognized as a clear and consistent policy based on values generally accepted in our part of the world.

Conclusion

If scientists take the approach of ethical dialogue, it will not directly affect opinion polls concerning the public perception of transgenic animals. However, indirectly the approach may have quite significant effects.

Firstly, it should be noted that journalists interested in animal biotechnology may be affected by the approach. Journalists often get their information by interviewing scientists. If the scientist is able to engage in an ethical dialogue with the journalist, he will stand a good chance of being quoted and he may affect the way the journalist approaches the whole issue.

Secondly, regulation of animal biotechnology is not decided mainly by opinion polls. Rather, members of ethical councils and committees, civil servants and politicians who take a special interest in the issue are the key persons; and they will often be both willing and able to engage in an ethical dialogue. Even if the views of the scientists are not fully accepted, they will be taken much more seriously if they are backed by ethical thinking which addresses the concerns of the public.

---

3 It should be noted that discomfort above a certain level may be unacceptable whatever may be the benefit of the experiment, e.g. as stated in Danish legislation on animal experimentation.

Finally, a major advantage is that the scientists themselves become more conscious about the ethical questions to which their work may give rise, and that they become more aware of what they think should be the ethical principles guiding research and the development of animal biotechnology. This may counteract a possible ethical stagnation in the scientific community and help to increase the public credibility of scientists, who are really serving vital interests by working with transgenic animals (Forsman and Sandøe 1995).

There seem to be only two possible disadvantages of this approach. The first is that the scientists will have to engage in ethical thinking. The other is that the ethical arguments may turn out to have quite significant effects on the ethical attitudes prevalent in the scientific community. However, ethical thinking may reveal that these alleged disadvantages are really to be counted as advantages.

Acknowledgements

The authors want to thank L.F.M. van Zutphen for useful comments to an earlier version of the paper.

References

Buehr M, Hjorth JP (1994) Genetically modified animals. Copenhagen: Danish Environmental Protection Agency.

Consensus Conference on Technological Animals, held 23-25 September 1992 at the Bella Centre in Copenhagen, Organized by the Danish Board of Technology in Cooperation with the Research Committee of the Danish Parliament, Final Document (English translation). Copenhagen: Danish Board of Technology.

Donelley S, McCarthy CR, Singleton R (1994) The brave new world of animal biotechnology. Hastings Center Report, January-February Issue.

Forsman B, Sandøe P (1995) Public participation in ethical decisions. In: Animals in Science Conference (Johnston NE, ed). Clayton, Victoria: Research Ethics Unit, Monash University, pp 20-25.

Fox M (1990) Transgenic animals: Ethical and animal welfare concerns. In: The Bio-Revolution (Wheale P, McNally R, eds). London: Pluto Press, pp 32-45.

Pursel VG, Pinkert CA, Miller KF, Bolt DJ, Campbell RG, Palmiter RD, Brinster RL, Hammer RE (1989) Genetic Engineering of Livestock. Science 244, 1281-1288.

Rollin B (1986) On telos and genetic manipulation. Between the species 2, 88-89.

Sandøe P, Holtug N (1993) Transgenic animals - which worries are ethically significant? Livestock Production Science 36, 113-116.

Sandøe P, Holtug N, Simonsen HB (1996) Ethical Limits to Domestication. Journal of Agricultural & Environmental Ethics, In press.

Sandøe P, Simonsen HB (1992) Assessing animal welfare - where does science end and philosophy begin? Animal Welfare 1, 257 -267.
Tait J (1990) NIMBY and NIABY: Public Perceptions of Biotechnology. In: The Bio-Revolution (Wheale P, McNally R, eds). London: Pluto Press, pp 224-236.
Vorstenbosch J (1993) The concept of integrity. Livestock Production Science 36, 109-112.

Legislation on Transgenic Animals in Europe

P. de Greeve and W. A. de Leeuw

Summary

Genetic modification of living organisms has found valuable applications in science, medicine, agriculture and industry. There are, however, few aspects of this new technology which need specific consideration, such as the regulation and control of animal welfare. In this overview, current regulations in Europe, dealing with transgenic animals, will be presented. Emphasis will be put on the welfare aspects.

On the basis of the information presented at a meeting in Strasbourg (September 1995) it seemed that, as far as the welfare aspects are concerned, no specific legislation is necessary since transgenic animals can be considered as 'normal' laboratory animals. Therefore, the creation, the breeding and the use of genetically modified vertebrate animals is subject to animal experimentation legislation when it concerns acts for experimental and/or other scientific purposes which could cause pain, suffering, harm or discomfort to the animal.

However, within the Council of Europe as well as the European Union, discussion continues as to whether genetic modification (of animals) should be regulated separately.

Introduction

Genetic modification of living organisms has found valuable applications in science, medicine, agriculture and industry. For example, in medicine, transgenic models have an enormous potential in the evaluation of new therapies. Also, the speed with which such intervention may be assessed is increased dramatically. However, some aspects of this new technology generate new problems for regulation, in particular finding the means to control genetic engineering in animal experimentation, with emphasis on animal welfare and also finding the means to ensure that releases of genetically-modified organisms into the environment will be safe.

For the last decade there has been growing public concern about the use of genetic modification and about the creation and production of transgenic animals. The Danish parliament has played an active role in the discussion and decision-making on biotechnology and gene technology. In 1987, as a result of parliamentary initiatives, Denmark was the first country in the world with a specific law on biotechnology. The two most important issues discussed in the parliamentary debate at that time were the risks for human beings and the risks for the environment from applications of gene technology. The genetic modification of animals was also discussed, but this was not given so much attention as the above mentioned two issues (Custers and Sterrenberg 1992). However, also in 1987, at a consensus conference, a panel consisting of lay people, concluded unanimously that genetic engineering on animals is unacceptable.

Five years later, the Consensus Conference on Technological Animals was organised in Denmark to discuss questions regarding ethical, health, economic, ecological and animal welfare aspects (TeknoliNaevnet 1992). With respect to the welfare of transgenic animals (theme 7), among others, two basic questions were addressed:

- What is the definition of animal welfare with regard to food production and other types of production in agriculture, bioreactors and laboratory animals? With respect to laboratory animals, the final document states that "it seems reasonable in some instances to eliminate general considerations on animal welfare".
- How much research is carried out in the area of animal welfare and what methods are being used? It was noted that one cannot directly measure whether an animal is suffering. Therefore, it was concluded that more research in the area of animal behaviour and physiology is necessary.

In the Netherlands, the birth of the transgenic bull, Herman, in 1993 was the impetus for extensive parliamentary consultation resulting in legislation on the justification of genetic modification of animals. In addition, in 1993, a public debate with a panel of lay people was organised dealing with the question "Genetic modification of animals, should it be allowed?" (NOTA 1993). Two salient outcomes of this debate were:

- the majority of the panel would like to see a temporary stop to the genetic modification of animals; and
- lay people should have a place on the review boards for genetic modification of animals.

At that time, various polls were conducted to find out how many people were in favour of the production of human medicine, produced by genetic modified animals. The Dutch market-research institute found that 11.6% of the citizens interviewed find it "absolutely unacceptable" that hereditary features of animals are changed in order to produce medicine for human diseases. About 44% found it "desirable" and about 19% believed it was "necessary".

The Regulations Dealing with Transgenic Animals in Europe

In 1976, the Council of Europe issued the Convention for the protection of animals kept for farming purposes. In 1992, after due consideration, a Protocol of Amendment to the Convention extended the scope of the Convention to the breeding of animals produced as a result of genetic modification or novel genetic combinations.

The European Union issued a number of regulations and directives that are of importance for the creation or use of genetically modified animals. The main goal of the directives is to harmonise national legislation; this is of great importance for the realisation of a true internal market. In chronological order the following regulations and directions on genetically modified animals have been issued:

* Draft proposal for a council regulation on the placing on the market of genetically modified animals and products thereof (VI/5108/88). This proposal on the sale of genetically modified animals and/or products originating from them is intended to harmonize national laws and to encourage consumer safety. The criteria used are: quality, efficacy and safety.

* Council directive (86/609/EEC) on the approximation of laws, regulations and administrative provisions of the Member States regarding the protection of animals used for experimental and other scientific purposes. This directive - adopted in 1986 - applies to all actions taken with animals for experimental and other scientific purposes. Animals are interpreted as: living non-human vertebrates, including free-living and/or reproducing larval forms, but excluding foetal or embryonic forms.

* Proposal for a council directive concerning the protection of animals kept for farming purposes (VI/5143/89). This proposal is based on the European Convention for the protection of farm animals kept for farming purposes. The criteria used in this proposal are: suffering, harm, health and welfare of the animals.

* Council regulation on the contained use of genetically modified micro-organisms (90/219/EEC). This regulation applies to all activities in which genetically modified micro-organisms are used, bred, stored, transported, destroyed or removed, with the use of physical barriers or a combination of physical, chemical and/or biological barriers in order to prevent genetically modified micro-organisms coming into contact with human beings or the environment. The criteria used in this regulation are: human health risks and risks to the environment.

* Council regulation (Directive 90/220/EEC) on the deliberate use of GMO's into the invironment. This regulation applies to all genetically modified organisms released deliberately into the environment without any form of containment, such as physical, chemical and/or biological barriers. The release of a genetically modified farm animal into a securely fenced meadow is not seen as a deliberate release of a genetically modified organism into the environment.

What about Legislation on the Welfare of Transgenic Animals in Europe ?
In Europe, two important documents have been issued with respect to the control of the use of animals in experiments.

The first one was initiated by the Council of Europe in Strasbourg. In 1985, after several years of discussion, the 26 countries of the Council of Europe agreed on the Convention for the Protection of Vertebrate Animals used for Experimental and other Scientific Purposes (ETS 123).

The second European document is the Directive for the Protection of Vertebrate Animals used for Experimental and other Scientific Purposes (86/609/EEC), which is mainly based on the Council of Europe Convention. However, the objectives of the two documents seem to be different: the Convention gives priority to important principles (animal rights, mankind's needs), whereas the main objective of the Directive is to harmonize national laws in order to avoid any distortions in the internal market.

In preparing the Convention and the Directive the consequences of genetic engineering for animal welfare were not taken into consideration, but when the discussion on the protection of transgenic animals started, the question was raised whether the creation and the breeding of transgenic animals are within the scope of the Convention and the Directive.

This issue has been discussed at length at the meetings of the Parties to the Conventions involved in Strasbourg and in Brussels, where the Competent Authorities (in relation to the Directive 86/609/EEC) meet regularly.

The Result of the Discussion in Brussels

With respect to the Directive 86/609/EEC it was concluded that transgenic animals to be used for experiments fall within the scope of this Directive. However, genetic modification is only covered by the Directive when there is some expectation that after birth the genetically modified animal will experience adverse effects on its health or welfare. The Harvard oncomouse could be used as an example here. This mouse is intentionally developed to be a very effective model in cancer research. The animal develops cancer very rapidly which causes suffering.

However, it must be stressed that genetic modification of animals is not automatically synonymous with suffering: if it is expected that the genetic modification will not result in any pain or distress for the born animal then it is not covered by the Directive.

The Result of the Strasbourg Meetings

The European Conventions contain a provision that parties should hold multilateral consultations to examine the progress of implementation of the Convention and the need for revision or extension of any of its provisions on the basis of new facts or developments. A conclusion of these consultations may have different forms which could have consequences for the text of the Convention itself (protocol of amendment) or give direction for the implementation of the Convention on national level.

* With respect to laboratory animals, on 27th November 1992, the Multilateral Consultation of Parties of the Convention ETS 123 adopted a resolution on the interpretation of certain provisions and terms of the convention. The Parties agreed "to understand the expression 'animals carrying harmful genetic modifications' as referring to genetically engineered animals and mutant animals capable of producing as a consequence offspring likely to suffer significantly". In other words, no distinction should be made between transgenic and other laboratory animals in the use of animals for experiments.

* Furthermore, with respect to the generation and breeding of transgenic animals, Parties agreed that "generating a transgenic strain is considered as constituting a procedure under Article 1, subparagraph 2 c of the Convention" and that "the breeding of animals carrying harmful genetic modifications may be considered on certain conditions to be determined by each Party as a procedure under Article 1, subparagraph 2 c. Such procedure must be carried out in accordance with the Convention." The last agreement especially can be considered as a compromise, because in some Member States the breeding itself is considered as a procedure.

More Information is Needed

On the basis of the above information, it can be concluded that the welfare of transgenic animals is covered by the Convention ETS 123 and the Directive 86/609/EEC, and that no additional legislation is needed. This view has been shared by the European Federation of Laboratory Animal Science Associations (FELASA), who has stated "that the use of transgenics in basic and applied biological medical and veterinary research is comparable with the use of naturally-occurring animal strains with spontaneous genetic abnormalities. If an animal suffers from a inherited disability, it is of no significance to that animal whether that disability arises from deliberate transgenic modification or is due to spontaneous genetic modification."

However, with respect to suffering of transgenic animals, it should be noted that one can never predict whether or not an animal will suffer as a result of genetic modification. Adverse effects can occur without notice, for instance due to insertional mutations.

The UK Home Office Inspectorate has stated that "for harm to be attributed it must be on the basis of clinical or observational grounds and it will be necessary to produce at least two generations of offspring before genetic modification can be regarded as harmless". However, with respect to knock-out mice, Dr. Martin van der Valk, a pathologist of the Netherlands Cancer Institute, argues that

> "this will not guarantee that there will be no suffering in the future, even after two generations: not the actual number of generations, but the rate of inbreeding (both with respect to the background genome and the hemi/hetero- and homozygosity of the transgene/knock-out gene) is important. On the other hand, one should note that several transgenes show no adverse effects unless combined with viral or chemical treatment, or in crosses with other transgenic animals. As such these transgenic animals do not significantly differ from conventional strains, including immune-suppressed/deprived ones (nude, SCID-like). Due to (theoretical?) possibilities in making transgene/knock out constructs, the occurrence of extremely rare or unknown 'diseases' in mice should be anticipated. This is also true with respect to the organ(s) affected: meticulous life-time observation and postmortem examination are regarded essential and beneficial for both animal welfare and science."

In this respect, I ask urgently for support for an initiative of the Multilateral Consultation at the Strasbourg meeting, September 1995. Parties are invited to inform the secretariat of the Council of Europe about any examples of harmful effects which have appeared in the offspring of transgenic animals. On the basis of the reported cases, combined with the results of the study to be performed by Drs. Miriam van der Meer at the Department of Laboratory Animal Science - Utrecht (see pages 87-88), insight could be gained into (potential) adverse affects of genetic engineering for the welfare of transgenic animals.

What about the application of the technique of genetic modification in respect of the concept of the 3Rs of Russell and Burch 1959), in particular the R of reduction of the number of animals used.

It has been argued, in particular by representatives of animal welfare associations, that bio-technology will increase the frequency of research on animals and thus will entail more suffering. However, Lathe and Mullins argued that

> "this is almost certainly not true in the case of human inherited disease. Over many years large numbers of animals have been employed in cystic fibrosis research, few of which are exceptionally good models and data so obtained have been of limited value. Transgenic models build upon earlier work while permitting accurate data using a very small number of animals and with confidence that the results correspond closely as possible to the human disease."

Nevertheless, there is sufficient ground for concern because the number of transgenic strains is increasing enormously and this will undoubtedly have consequences for the welfare of animals. Therefore, according to the objectives of the Convention and the Directive, Member States should identify criteria for guidelines on creation and breeding of 'animals carrying harmful genetic modifications' with specific attention on welfare aspects. This has been particularly emphasized by the Multilateral Consultation of the Parties to the Convention ETS 123 of the Council of Europe.

Recent Developments

In October 1995, a working party was organized in Strasbourg to prepare for the next Multilateral Consultation of the Parties to the Convention ETS 123. The Parties and the other Member States of the Council of Europe were requested to provide information on any regulations dealing with the protection of transgenic animals. On the basis of the information presented it became clear

- that no specific legislative provision with respect to the welfare of transgenic animals exist in any of the Parties;

- that the creation, the breeding and the use of genetically modified vertebrate animals fall under animal experimentation legislation if it concerns acts for experimental and/or other scientific purposes which could cause pain, suffering, harm, discomfort to the animal;
- that in a few Member States, the public seems to be relatively well-informed about biotechnology;
- that in other Member States the discussion has recently started and/or advisory boards have been established;
- that in several Member States the Directives 219 and 220 have been implemented into national legislation; and
- that in a few Member States there is no political and public discussion on genetically modified animals nor about the ethical aspects involved.

On the basis of a contribution of FELASA and the discussion in Strasbourg, a statement was drafted and approved by the representatives of the Parties and other Member States, which clearly illustrates the current situation. It was agreed that the generation and maintenance of a transgenic strain must always call for careful consideration, especially when there is a risk of deleterious characteristics. If the actual welfare problems of transgenic animals were considered different than those that apply to laboratory animals in general, it was acknowledged that there is a chance of a higher incidence of unpredictable adverse effects. Therefore, the need for meticulous observation of the animals throughout the entire lifespan and possibly for more than one generation was emphasized.

Finally, it was concluded that, considering the speed of developments in transgenic technology, and the possible applications of these techniques in particular in the field of biomedical research, ethical issues will continue to arise for consideration, which would need to be considered in particular in terms of justification, accountability and competence.

Is there a Need for a New Legal Instrument?

On the basis of the above information, it could be concluded that there is no need for a special regulation with respect to the welfare of transgenic animals. However, due to the ethical and also the economic and legal aspects involved and to the increasing concerns of the public at large, the question whether genetic modification in general requires specific regulation has been given great consideration by the European Union and the Council of Europe.

On 21 June 1995, the EU-Group of advisers on the ethical implications of biotechnology met in Brussels. According to the minutes of the meeting some of the participants "considered that it would be sufficient to adapt existing directives to cover transgenic. Others felt that the significant changes brought about by the genetic modification of animals generated a need for special legislation."

At the Council of Europe, the Parliamentary Assembly in its Recommendation 1213 on developments in biotechnology and the consequences for agriculture, called for the elaboration of a new legal instrument covering the field of biotechnology including transgenic animals. Replying to this Recommendation and in the light of the opinions communicated by two committees competent respectively in the field of animal protection and bio-ethics, the Committee of Ministers on 7-11 September 1995 agreed, to envisage holding of a "conference with representatives of all relevant professions and interest groups, in particular representatives of the world of science and ecology, law, industry, consumers, the health field and protection of animals". The committee also wishes to associate the European Union and the European Patent Office with such an conference. The conclusion of this conference, which would be preceded by studies of the various aspects involved, could prepare the way for determining the need to draw up a new Convention covering bio-ethical aspects of biotechnology.

I totally agree with this approach and I am convinced that only a combined effort by all the organizations and people involved can contribute to sensible decision-making. And most importantly, this decision illustrates the change of mind that has taken place in the way animal welfare is considered. Indeed, this is not considered as an isolated issue any more, but it is an integral part of the ethical approach to a problem. The problems raised by the developments of biotechnology require a multidisciplinary approach.

Addendum

In The Netherlands, with respect to the legislation on transgenic animals, it has been decided that proposals for projects including the development of transgenic animals should be evaluated by three separate Committees:

1. The Committee on Genetic Modification (COGEM) advises the Minister of Housing, Special Planning and Environment on the safety and risk aspects for the environment.
2. Within the framework of the Animal Health and Welfare Act (1992) a system of licensing will be developed with respect to experiments aiming at the genetic modification of animals. A national committee on ethical evaluation of genetic modification of animals, called the Committee on Biotechnology, will advise the Minister of Agriculture, Nature Manage-

ment and Fisheries on the ethical aspects of the creation and the use of transgenic animals in general and on the admissibility of proposed projects.

3. According to the revised Experiments on Animals Act, which is likely to come into force at the beginning in 1996, any experiment on animals shall be prohibited unless
   a. a positive recommendation has been given by a local Ethical Review Committee, which has been officially recognised by the Minister of Health, Welfare and Sport, or
   b. a negative recommendation given by the Ethical Review Committee, which is followed by a positive decision on the part of the national Animal Experiments Advisory Committee.

It will be obvious that the detailed mechanisms of these three types of evaluation need due consideration. This is especially so with regard to the evaluation by the local Ethical Review Committees and the national Committee on Biotechnology where some overlap and interference may occur.

Some people have the opinion that the ethical aspects concerning the development, the breeding and the use of transgenic animals are not fully covered by the evaluation of the Ethical Review Committee. However, on the other hand, the national Committee on Biotechnology will not consider in the first place the detailed aspects of individual experiments (number of animals used in the experiment, level of distress of the animals used in relation to the aim of the experiment). There is still discussion on the question of whether or not a general exemption must be given from the evaluation by the national Committee on Biotechnology for small laboratory animals used for basic biomedical research.

It is clear that one is still struggling with the problem of how to set up an efficient evaluation system based on three different legal frameworks to consider fully ethical, welfare and safety aspects, in such a way that duplications are prevented and that scientists are not bothered disproportionately by an abundance of administrative requirements.

Acknowledgement

The authors wish to thank Mrs. L. Lwoff, Council of Europe, for her useful suggestions.

References

Custers R, Sterrenberg L (1992) Regulation and discussions on genetic modification. NOTA, the Netherlands Organization for Technology Assessment (the Rathenau Institute). P.O Box 85525, 2808 CE The Hague, the Netherlands.

Lathe R, Mullins JJ (1993) Transgenic animals as models for human disease - report of an EC Study Group. Transgenic Research 2, 286-299.

NOTA (1993) Public debate genetic modification of animals, should it be allowed? NOTA, the Netherlands Organization for Technology Assessment (the Rathenau Institute). P.O Box 85525, 2808 CE The Hague, the Netherlands.

Russell WMS, Burch RL (1959) The Principles of Humane Experimental Technique. London: Methuen.

TeknoliNaevnet (1992) Conclusions of the Question Panel at the Consensus Conference on Technological Animals 23-25 September 1992, Copenhagen, Denmark.

Van der Valk M (1995) personal communication.

The Utrecht Workshop on Welfare Aspects of Transgenic Animals

General Discussion

T. Poole

The main points raised in the discussion will be briefly outlined. Firstly, some delegates asserted that there is no real difference between transgenics and genetical changes achieved through traditional selective breeding. The opponents of this view argued that the changes brought about by transgenics are less predictable, much more rapid and in some cases much more debilitating. The point was made that genes from different species may have been separated for millions of years and that their insertion into the genome could result in a serious upset to the "genetic integrity" of the species. The concept of genetic integrity was debated with some delegates supporting such a notion and others denying its existence. The main arguments against genetic integrity are that, in the course of evolution, genetical differences between species or sub-species are often non-adaptive, having arisen through physical separation of sections of an originally single population and that in some instances hybrids may even be at an advantage. Furthermore transgenesis by means of viral vectors is a naturally occurring phenomenon.

The concept of the scale of welfare put forward by Broom was discussed. An interesting discussion developed on the question whether the welfare of an immuno-suppressed animal in an SPF environment, where it has never been challenged by an infection, is still compromised. In further discussion, there was general agreement that welfare cannot be measured directly because it refers to a subjective state of mind. However, welfare can be assessed by using behavioural, physiological and pathological signs. The approach, as presented by Van der Meer and Van Zutphen in proposing to develop a model protocol for assessing the welfare of transgenic animals, is based on this concept. Poole proposed, as an alternative to the welfare scale, one which symbolised the state of mind of the animal from pleasure at one end to extreme pain or distress at the other (see Table). Concerns for the welfare of some transgenic animals result from the fact that individuals that are not the subject of experiments (stock and breeding animals) may spend their whole lives in a state in which they are incapable of coping with the suffering generated by their genetic condition.

The question was raised as to whether animals could be bred which are contented to exist in deprived environments. This might further be elaborated to ones which feel no pain and cannot suffer. The ethical aspects of breeding what might effectively be non-sentient animals certainly would be of considerable complexity and create very considerable concern among animal welfarists and philosophers.

Finally, delegates agreed that the workshop had provided excellent opportunities for an exchange of information on both the scientific and welfare aspects of transgenic animals. It was hoped that a further meeting would be held at a future date further to resolve some of the complex issues which had been raised.

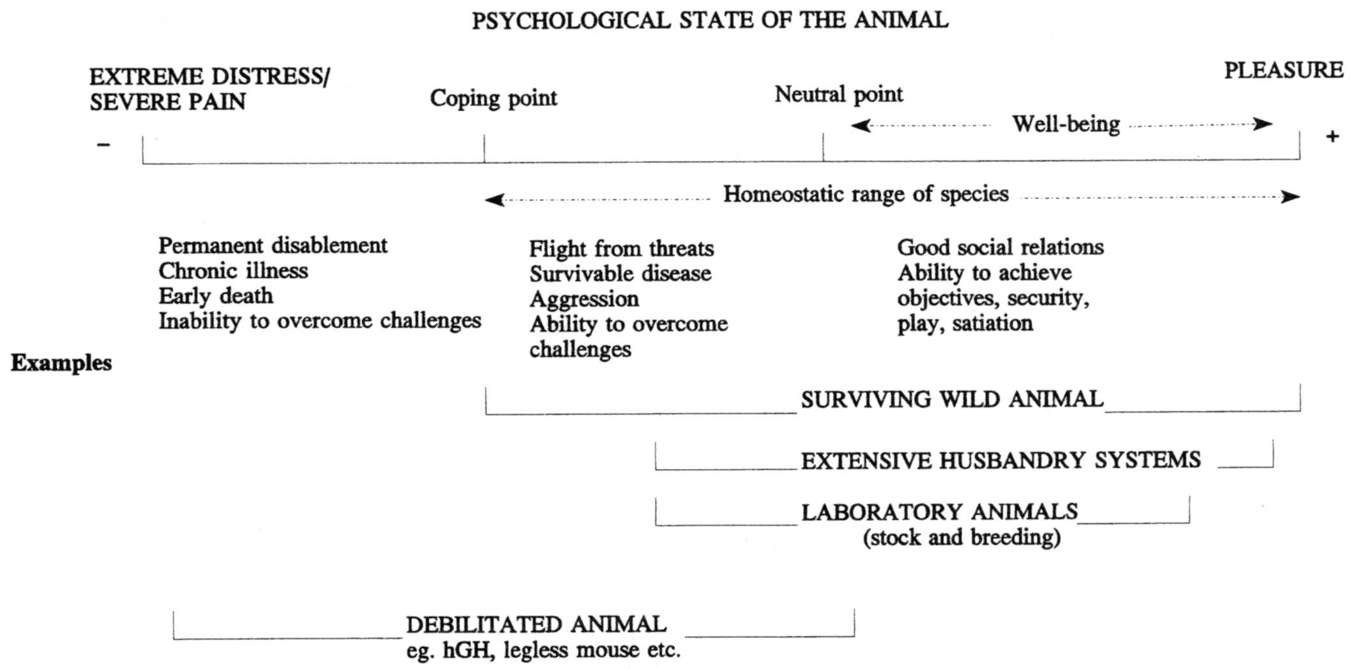

Table: *Animal welfare is a psychological concept associated with the animal's state of mind ranging from extreme distress to pleasure. On this scale there will be a neutral point where the animal is neither suffering nor experiencing pleasure. Wild animals will experience a range of feelings either side of the neutral point. However, there is a further negative scale where the animal cannot cope and suffers chronic illness or death. Captive, non-experimental stock will generally experience less of the negative part of the range than wild animals because they are protected from predators, illness and injurious aggression from conspecifics. The concern is that some transgenic animals live in a range which is outside that within which an animal can cope.*

List of Participants

Dr. V. Baumans, Dept. of Lab. Anim. Science, Utrecht University, Utrecht, The Netherlands

Dr. R. ten Berg, Netherlands Cancer Institute, Amsterdam, The Netherlands

Prof.dr. G.M.A. van Beynum, Gene Pharming, Leiden, The Netherlands

Dr. E. Bleibaum, Commission of the European Communities, Brussels, Belgium

Dr.ir. H.J. Blokhuis, ID/DLO, Lelystad, The Netherlands

Drs. F. Brom, Centre for Bioethics and Health Law, Utrecht University, Utrecht, The Netherlands

Prof. D.M. Broom, Dept. Clin. Vet. Medicine, Univ. of Cambridge, Cambridge, United Kingdom

Dr. D. Bruce, Church of Scotland, Society, Religion and Techn. Proj., Edinburgh, Scotland

Dr. K. Bürki, Preclinical Research, Sandoz Pharma Ltd., Basel, Switzerland

Dr. P. Costa, Inst. di Ricerche di Biol. Molecolare, Rome, Italy

Prof.dr. Tj. de Cock Buning, University of Leiden, Leiden, The Netherlands

Ir. R.T.J. Custers, COGEM (Commission Genetic Modification), Utrecht, The Netherlands

V. Delpire MSc, ECVAM (European Centre for the Validation of Alternative Methods), Ispra, Italy

Prof.dr. J.E. van Dijk, Dept. of Pathology, Utrecht University, Utrecht, The Netherlands

Dr. L. Fretz, Technical University Delft, Delft, The Netherlands

Drs. P. de Greeve, Department of Animal Experimentation, Min. VWS, Rijswijk, The Netherlands

Drs. W. van Haeringen, Dept. of Lab. Anim. Science, Utrecht University, Utrecht, The Netherlands

Drs. H. van Herck, Centr. Lab. Anim. Inst. (GDL), Utrecht University, Utrecht, The Netherlands

Dr. M. Jennings, RSPCA, Horsham, West Sussex, England

Dr. A. Klepsch, Commission of the European Communities, Brussels, Belgium

Drs. W. de Leeuw, Department of Animal Experimentation, Min. VWS, Rijswijk, The Netherlands

Drs. M. Linskens, Dutch Society for the Protection of Animals, The Hague, The Netherlands

Mr.drs. H. Lommers, Raad voor Dieraangelegenheden, The Hague, The Netherlands

Drs. M. van der Meer, Dept. of Lab. Anim. Science, Utrecht University, Utrecht, The Netherlands

Dr. B. Mepham, Centre for Appl. Bioethics, Univ. of Nottingham, Leicestershire, England

C. Moore, MSc, Centre for Appl. Bioethics, Univ. of Nottingham, Leicestershire, England

Dr. W. Müller, Institute for Genetics, University of Cologne, Cologne, Germany

Drs. A. Nap, Min. of Agriculture, Nature Management & Fisheries, The Hague, The Netherlands

Mr.ir. A. van der Neut, Min. of Agriculture, Nature Management & Fisheries, The Hague, The Netherlands

Dr. T. Poole, UFAW (Universities Federation for Animal Welfare), Potters Bar, Herts, England

Ir. C.G. van Reenen, ID/DLO, Lelystad, The Netherlands

Dr. P. Sandøe, Bioethical Research Group, University of Copenhagen, Copenhagen, Denmark

Prof.dr. E. Schroten, Centre for Bioethics and Health Law, Utrecht Univ., Utrecht, The Netherlands

Dr. F. Stafleu, Centre for Bioeth. and Health Law, Utrecht Univ., Utrecht, The Netherlands

Dr. E. van Tilburgh, Veterinary Dept., Ministry of Agriculture, Brussels, Belgium

Dr. J.S. Verbeek, Dept. of Immunology, University Hospital, Utrecht, The Netherlands

Dr. J. Vorstenbosch, Centre for Bioethics and Health Law, Utrecht Univ., Utrecht, The Netherlands

Prof.dr. J.G. Vos, Lab. of Pathology, RIVM, Bilthoven, The Netherlands

Dr. D.B. Wilkins, Eurogroup for Animal Welfare, Brussels, Belgium

Prof.dr. E. Wolf, Mol. Tierzucht und Haustiergenetik, Ludwig-Maximilians-Universität München, München, Germany

Prof.dr. L.F.M. van Zutphen. Dept. of Lab. Anim. Sci., Utrecht Univ., Utrecht, The Netherlands

Index

Springer and the environment

GPSR Compliance

The European Union's (EU) General Product Safety Regulation (GPSR) is a set of rules that requires consumer products to be safe and our obligations to ensure this.

If you have any concerns about our products, you can contact us on ProductSafety@springernature.com

In case Publisher is established outside the EU, the EU authorized representative is:

Springer Nature Customer Service Center GmbH
Europaplatz 3
69115 Heidelberg, Germany

Batch number: 10844675

Printed by Printforce, the Netherlands